Statistics

SIMPLIFIED
AND SELF-TAUGHT

Statistics

SIMPLIFIED
AND SELF - TAUGHT

Stanley H. Stern
Adjunct Professor, Management
Antioch University
San Francisco, CA

ARCO PUBLISHING, INC.
NEW YORK

Published by Arco Publishing, Inc.
215 Park Avenue South, New York, N.Y. 10003

Library of Congress Cataloging in Publication Data

Stern, Stanley H.
   Statistics simplified and self-taught.

   1. Statistics.   I. Title.
QA276.12.S74 1984   519.5   83-25762
ISBN 0-668-05813-7 (Paper Edition)

Printed in the United States of America

10 9 8 7 6 5 4 3 2 1

CONTENTS

INTRODUCTION
"The Nature of the Beast"

Why is it that the average person cringes at the mere mention of the terms statistics and probability?

Why is it that the very thought of taking a course called statistics is enough to turn an "A" student into a neurotic mass?

What is an "A" student?

Who or what is the average person?

Who or what is a Dow Jones Average?

There is a long-held view that you can prove any statement of fact using statistics. That is an extreme notion, but a knowledge of statistics and probability can certainly increase your understanding of many of the questions and problems confounding people and institutions in the present-day world. Not everyone today has to be a statistician, but everyone, like it or not, must be a consumer of statistics and probability.

Statistics is the mathematical method of dealing with the collecting, presenting, analyzing, and utilizing of numerical facts and figures in order to draw conclusions and aid in making decisions in an uncertain world— business, politics, medicine, and, with the use of computers, romance.

Probability is the mathematical theory based on a process developed by the French mathematician Blaise Pascal (circa 1654) upon which Las Vegas and Atlantic City casinos are built and flourish. Probability theory employed on selected random events (i.e., tossing a coin) will tell us who should win, if it might rain, or if we should buy more insurance.

To that multitude of citizens who break out in a cold sweat, or those cowards who dismiss the entire subject as too difficult or a waste of their time, I say, "Hogwash!" The visions of arithmetic formulas piled high atop mountains of numbers in threatening arrays of tables; the language as foreign as Greek; the reams of meaningless facts and figures to decipher—tax programs, mortality rates, census data, the life expectancy of light bulbs; and the demographics of an FM classical music station—these misconceptions of a discipline that seems more a quagmire than a fountain of practical, logical information will be dispelled posthaste.

TRUST ME!

I will demystify the study of statistics for all you nonbelievers. I will defang the monster and render it harmless and useful. Statistics is not the personal instructional tool of any one science or discipline. Every segment of our empirical world can benefit from the employment of a set of statistical methods. *You* can benefit from an understanding of the essential principles and practices of statistics and probability. The objective of this guide is not to create an army of statisticians. I am a teacher who wants the reader, the student, to acquire statistical techniques and insight to employ in a world that in the last quarter of the twentieth century is dependent on an invaluable set of procedures to research workers in every quantitative and qualitative discipline:

a. Market research—New products and test markets.
b. Insurance—Life, health, and property.
c. Politics—Federal, state, and local voting patterns.
d. Taxes—Federal, state, and local.
e. Psychology—To compare the effectiveness of various therapies.
f. Medical research—What's killing us and why?
g. Sports—How the teams and players match up.
h. Gambling—Roulet, craps, poker, and the horses.

So don't be an ostrich!

Statistics will not go away. To the statistician, statistics is the means to the end. To the reader, whether student or interested layperson, statistics is a tool for survival and greater success in a complex society where generalizations are made hazardous by the quantity of variables and the limitations of field experimentation. Scientific statements are only meaningful if they lead to observations which can be constructed on a world delineated by our five senses. Statistics is a powerful tool for understanding critical relationships in complicated situations. This is particularly true in the sciences that deal with human beings—biology, medicine, and the social sciences.

Do you happen to suffer from math anxiety or math avoidance? Many readers bring a built-in prejudice to math-related fields of study. For a few fortunate souls, mathematical acumen is a gift from the gods. For the rest of the world (this writer included) mathematics, through intermediate algebra, remains an amalgamation of learned (and often forgotten) numerical skills. This guide through the fields of descriptive and inferential statistics can and will allay the fears of those with the little and big doubts that plague the millions of average and above-average men and women who suffer from "statisticsaphobia."

Now, with your fears under control (or at the very least in abatement) and armed with a good supply of sharpened pencils, you are ready to slay those statistical dragons. And don't forget, I'll be with you every step of the way.

1

THE LEXICON
"Statistics to English"

1. **Descriptive Statistics:** The methods of collecting, organizing, analyzing, and utilizing numerical data derived from the empirical world.
2. **Inferential Statistics:** The methods used to describe a population (universe) by studying a random sample of that population.
3. **Data:** Facts and figures collected on some characteristic of a population or sample.
4. **Population or Universe:** All the members of a particular group of items or individuals.
5. **Sample:** Any subset of a population.
6. **Parameter:** One characteristic of a population (a descriptive measure).
7. **Statistic:** One characteristic of a sample.
8. **Variable:** A measure or characteristic that may have a number of different values.
9. **Discrete Variable:** A characteristic which can only assume designated values (i.e., people on line, number of accidents). How many . . . ?
10. **Continuous Variable:** A characteristic that can assume any value on the measurement scale employed (i.e., weight, height, temperature). How much . . . ?
11. **Quantitative Variable:** A measure or characteristic with an exact numerical value (i.e., number grade on an exam).
12. **Qualitative Variable:** A measure or characteristic with an inexact numerical value, e.g., a letter grade on an exam.
13. **Frequency:** The number of times a score or a group of scores (class) occurs in a population or sample.
14. **Class:** A group of scores in a population or sample.
15. **Class Mark:** Halfway between the upper limit and the lower limit of a class (midpoint).
16. **Relative Frequency:** The frequency of one score or group of scores divided by the total frequency of all the observations.
17. **Cumulative Frequency:** The frequency of any class plus the frequencies of all preceding classes in a distribution.
18. **Class Boundary:** Halfway between the lower limit of one class and the upper limit of the preceding class (The exact limit).
19. **Frequency Distribution:** A tabulation of data showing the number of times a score or group of scores appears.
20. **Graph:** A pictorial representation of a set of data.
21. **Histogram:** A vertical bar graph that shows the frequencies of scores or classes of scores by the height of the bars.

22. **Frequency Polygon:** A graph on which the frequencies of classes are plotted at the class mark and the class marks are connected by straight lines.
23. **Ogive:** A graph of cumulative frequency distribution plotted at the class marks and connected by straight lines.
24. **Scattergram:** The relationship between two variables is shown by a series of dots plotted on a graph.
25. **Central Tendency:** A central value between the upper and lower limits of a distribution around which the scores are distributed.
26. **Mean:** The arithmetic average of all the scores or groups of scores in a distribution.
27. **Median:** A point in the distribution of scores at which 50 percent of the scores fall below and 50 percent of the scores fall above.
28. **Mode:** The most frequently appearing score or group of scores appearing in a distribution.
29. **Sigma Notation (Σ):** The Greek capital letter sigma indicates addition of whatever follows immediately in the mathematical expression.
30. **Range:** The highest score in a distribution minus the lowest score.
31. **Deviation:** The difference between a score or the class mark of a group of scores and the arithmetic mean.
32. **Average Deviation:** The sum of the differences between scores or class marks and the arithmetic mean divided by the total frequency of the sample.
33. **Variance:** A measure of the dispersion of scores in a distribution away from the arithmetic mean. The mean of the squared deviations about the mean.
34. **Standard Deviation:** A quantitative measure defining the extent to which scores are dispersed throughout the distribution in relation to the arithmetic mean. The square root of the variance.
35. **Probability:** The relative frequency of a random event occurring. The number of favorable outcomes divided by the total amount of outcomes.
36. **Random Sample:** A subset of a population in which all outcomes have the same probability of occurring and in which each outcome's selection is independent of the selection of all other outcomes.
37. **Dependent Event:** An event whose probability of occurrence is based upon the occurrence of another event (conditional probability).
38. **Independent Event:** An event where the probability of occurrence is not based upon the occurrence of another event.
39. **Venn Diagram:** A pictoral description of the probability concepts of independent and dependent events.
40. **Probability Distribution:** A tabulation of all probabilities for all the possible outcomes of a particular random event.
41. **Sampling Distribution of the Mean:** Based on a number of equal-sized random samples of a population, the mean of a sampling distribution will be the mean of the population.

42. **Normal Curve:** A continuous, unimodal, symmetrical, bell-shaped curve with the maximum height at the mean. The standard normal curve has a mean equal to zero (0) and the standard deviation of one (1).

43. **Standard Error of the Mean:** The standard deviation of the sampling distribution of the mean.

44. **Z Score:** A standard score which indicates the number of standard deviations a complementary raw score is below or above the mean.

45. **T Score:** A standard score with a mean of fifty (50) and a standard deviation of ten (10) used to convert raw test scores.

46. **Central Limit Theorem:** The mathematical basis for using the normal distribution as the sampling distribution of all the sample means from an established sample size.

47. **Estimator:** A statistic relevant for estimating the parameter.

48. **Confidence Interval:** A range of values believed to contain the population characteristic.

49. **Statistical Hypothesis:** A statement, declaration, or claim about the nature of a population.

50. **Null Hypothesis:** A statement or assertion that is being tested for rejection.

51. **Test Statistic:** A statistic used in carrying out the test of a hypothesis.

2

ARRANGING THE DATA
"The Collection and Presentation of the Data in Pretty Pictures"

Research results in the accumulation of numbers that represent the scores of the conditions observed. Researchers, students, and the varied professionals are the collectors of numerical bits of information who seek out methods to present in ordered form the masses of numbers they collect. The statistical methods this book will present are merely the tools for bringing order into accumulated data and to aid students and professionals in making reasonable decisions in an uncertain world. The arranging of data in a specific, given order will provide the structure for larger, more expansive concepts.

FREQUENCY DISTRIBUTIONS

The statistician's fundamental technique for putting into useful order a disarray of collected data is the frequency distribution. To put it simply, the frequency distribution is a tried and true systematical procedure for displaying scores from low to high in relation to a selected quantifiable characteristic (parameter).

If test scores are being observed, the scores are recorded in a column with the highest scores recorded first. This is called "rank distribution." Since it is likely that more than one individual will receive the same test score, a second column is needed to show how many people got a specific score or group of scores (class). Such an arrangement of the collected data, test scores in this example, offers the researcher insight into their nature and significance.

For example, a university gives a standardized entrance exam consisting of 100 math questions to 100 entering freshmen. The academic ability scores for the students are the data which is collected by the administration to determine placement.

Step 1: To organize these 100 scores shown in Table 2-A, determine the "range."

Table 2-A. The Scores of 100 Entering Freshmen on a Standardized Test

73	29	82	71	68	59	91	63	84	47
69	53	81	53	39	17	82	70	80	59
82	37	70	47	48	29	36	49	91	23
34	22	69	61	59	38	59	78	76	61
63	69	68	82	77	81	83	43	58	70
71	84	57	79	82	63	72	83	61	57
56	90	53	83	71	49	59	83	68	72
55	62	71	24	65	17	63	43	49	68
49	59	41	97	49	43	81	72	63	70
77	42	30	64	74	62	90	88	53	88

The range is the highest score in the distribution minus the lowest score. For the 100 scores in Table 2-A, the range = 97 − 17 = 80.

Step 2: To organize and present the scores in a practical way, a frequency distribution table is developed. The tabulation for this presentation consists of keeping the scores in ascending order in column 1 of the frequency distribution. Thus, the scores (data) will be classified into as many categories as there are scores. However, with large samples of scores, it is generally impractical and time-wasting to retain as many categories as there are scores. For example, in Table 2-A, if no scores are repeated, each having a frequency of one, then we would need 100 categories for column 1. So to make the tabulation less unwieldly, we reduce the number of categories into groups or *classes* of data. However, such reclassification does result in the loss of certain information about the actual scores in the samples. For example, the frequency distribution in Table 2-B no longer contains the scores 41, 71, 90, and so on. We learn instead there are 13 scores in the class 41-50.

Table 2-B. Columns 1, 2, and 3 of a Frequency Distribution Using Classes to Categorize the Test Scores

Class	Tally	Frequency
11-20	‖	2
21-30	ЖН I	6
31-40	ЖН	5
41-50	ЖН ЖН ‖‖	13
51-60	ЖН ЖН ЖН	15
61-70	ЖН ЖН ЖН ЖН ‖‖	23
71-80	ЖН ЖН ЖН	15
81-90	ЖН ЖН ЖН ‖‖	18
91-100	‖‖	3

When classes are used to represent groups of scores, the researcher must make two assumptions in order to present the frequency distribution graphically, as well as to compute the measures of central tendency and dispersion (*see* Chapter 3). The primary assumption is to use one score to represent the class of scores. The second assumption is that for any class, the scores grouped within are distributed uniformly between the lower and upper limits.

To determine the correct number of classes to employ in a frequency distribution, the accepted rule of thumb is to use between 8 and 15 classes to cover the total range of scores.

NOTE: Classes do not have to be of the same length. Two intervals with a frequency of one (1), for example, can be combined to form one larger class interval with a frequency of 2.

Step 3: After the classes are listed in ascending order in column 1 (*see* Table 2-B), a *tally* of the scores pertaining to each class is done in column 2. For example in Table 2-B, the researcher counted 15 scores in the class with the limits 51-60. The total of column 2 must equal the number of the scores in the distribution.

Step 4: The *relative frequency* of each score or group of scores is derived by dividing the frequency of a score or group of scores by the total frequency of the distribution as in column 4, Table 2-C.

Table 2-C. Frequency Distribution of 100 Test Scores Including Relative Frequency and Percentage of Frequency

Classes	Tally	Frequency	Relative Frequency	Percentage of Frequency
11-20		2	2/100 = .02	2%
21-30		6	6/100 = .06	6%
31-40		5	5/100 = .05	5%
41-50		13	13/100 = .13	13%
51-60		15	15/100 = .15	15%
61-70		23	23/100 = .23	23%
71-80		15	15/100 = .15	15%
81-90		18	18/100 = .18	18%
91-100		3	3/100 = .03	3%
TOTAL:	100	100	100/100 = 1.0	100%

Step 5: The *percentage of frequency* of each score or group of scores is obtained by translating relative frequency to a percentage by multiplying the relative frequency by 100 (that's by moving the decimal point two places to the right or by punching the percent key on your calculator) as in Table 2-C, column 5.

Step 6: The *class mark* (midpoint) is obtained as in Table 2-D, column 6, by adding the lower limit and the upper limit of a class interval and dividing by two. The class mark will serve as the representative value of the class for future computation.

Table 2-D. A Complete Frequency Distribution for 100 Scores of Academic Ability Presented to Entering Freshman Students

Class	Fre-quency	Relative Fre-quency	Per-centage Fre-quency	Class Mark	Class Bound-aries	Cumu-lative Fre-quency
					10.5	
11-20	2	.02	2%	15.5		2
					20.5	
21-30	6	.06	6%	25.5		8
					30.5	
31-40	5	.05	5%	35.5		13
					40.5	
41-50	13	.13	13%	45.5		26
					50.5	
51-60	15	.15	15%	55.5		41
					60.5	
61-70	23	.23	23%	65.5		64
					70.5	
71-80	15	.15	15%	75.5		79
					80.5	
81-90	18	.18	18%	85.5		97
					90.5	
91-100	33	.03	3%	95.5		100
					100.5	
TOTALS:	100	1.00	100%			

Step 8: The *cumulative frequency*, as shown in Table 2-D, column 7, is the class intervals. They are points that delineate the dividing points between two successive classes.

$$\text{Class Boundary} = \frac{\left(\begin{array}{c}\text{Upper Limit of} \\ \text{One Class}\end{array} + \begin{array}{c}\text{Lower Limit of the} \\ \text{Next Highest Class}\end{array}\right)}{2}$$

The class boundary is the point where, so to speak, one class of a discrete variable (i.e., whole numbers) leaves off and the next higher class takes over. The point must fall between two scores of a distribution. The lowest class boundary and the highest, as in Table 2-D, column 6, 10.5 and 100.5 give each class two boundaries.

Step 8: The *cumulative frequency,* as shown in Table 2-D, column 7, is used extensively in describing a distribution of scores. It is computed by adding the frequency of scores in a class interval to the frequencies of all preceding class intervals in the distribution. Just think of the class interval as pancakes that are placed in a progressively higher stack.

That, my friends, is the extent of a frequency distribution. In the following section, the information compiled and tabulated for the distribution will be presented in pictorial form (charts and graphs).

EXERCISES

Exercise 2-1. A math class of 30 students took an algebra test. The following scores resulted:

50 96 80 74 76 68 60 78 58 62 84 90 72 62 76
60 54 60 82 72 58 82 66 58 64 56 66 70 58 52

a) What is the range?
b) Complete the following frequency distribution of test scores:

Score	Frequency	Relative Frequency		Percentage Frequency
50	1	1/30 = .033		3.3%
_____	_____	_____	_____	_____ %
_____	_____	_____	_____	_____ %
_____	_____	_____	_____	_____ %
_____	_____	_____	_____	_____ %
_____	_____	_____	_____	_____ %
_____	_____	_____	_____	_____ %
_____	_____	_____	_____	_____ %
_____	_____	_____	_____	_____ %
_____	_____	_____	_____	_____ %
_____	_____	_____	_____	_____ %
_____	_____	_____	_____	_____ %
_____	_____	_____	_____	_____ %
_____	_____	_____	_____	_____ %
_____	_____	_____	_____	_____ %
_____	_____	_____	_____	_____ %
_____	_____	_____	_____	_____ %
_____	_____	_____	_____	_____ %
_____	_____	_____	_____	_____ %
_____	_____	_____	_____	_____ %
_____	_____	_____	_____	_____ %
_____	_____	_____	_____	_____ %
_____	_____	_____	_____	_____ %

c) Complete a frequency distribution of the 30 test scores using 10 classes.

Exercise 2-2. The data on rainfall (in inches) in a region of the United States are grouped into class intervals 11-15, 16-20, 21-25, 26-30, 31-35, and 36-40.

a) Find the class marks.
b) Find the class boundaries.

Exercise 2-3. Complete the following distribution of 150 men:

Class	Fre-quency	Relative Frequency	Per-centage Frequency	Class Mark	Class Bound-aries	Cumu-lative Fre-quency
125-139	10	10/150 = .067	6.7%	132		10
					139.5	
140-154	20					30
155-169	24					
170-184	26					
185-199	28					
200-214	17					
215-229	14					
230-244	11					
TOTAL:	150	150/150 = 1.00	100.0%			

CHARTS AND GRAPHS

In this section, we will consider the pictorial or graphical presentation of data. Specifically, we will take a serious look at the pie chart, the histogram, the frequency polygon, the cumulative frequency polygon, and the scattergram. The data for the pictorial presentations is derived from the tabulations in a frequency distribution like that in Table 2-D. Pie charts and graphs can be used to illustrate the relationship between the independent variable (number of students) and the dependent variable (test scores).

Each and every one of us comes across graphs and charts in our daily lives. The word "graph" is derived from the Greek work which means: "to be drawn or written." In our trip through the essential world of statistics, the term "graph" will be used to define any pictorial representation of a set of data.

The most common examples of pie charts and graphs appear in news-papers and magazines to illustrate government income and expenditures (pie chart), changes in interest rates (frequency polygon), annual rainfall (histogram), public school attendance (pictures of little children at their desks), and the consumption of alcoholic beverages (cumulative frequency polygon).

Pie Chart

The name pie chart is derived from the circular shape, which consists of slices cut out of the circle to illustrate what percentages of the total are ac-counted for by each individual score or group of scores. Pie charts are most effective when illustrating qualitative variables involving popularity.

NOTE: There are 360 degrees in a circle. Each class interval represents a percentage of 360 degrees.

Class	Percentage Frequency	Percentage of Pie Chart (Degrees)
11-20	2%	2% × 360 = 7.2
21-30	6%	6% × 360 = 21.6
31-40	5%	5% × 360 = 18.0
41-50	13%	13% × 360 = 46.8
51-60	15%	15% × 360 = 54.0
61-70	23%	23% × 360 = 82.8
71-80	15%	15% × 360 = 54.0
81-90	18%	18% × 360 = 64.8
91-100	3%	3% × 360 = 10.8
TOTAL:	100%	100% × 360 = 360.0

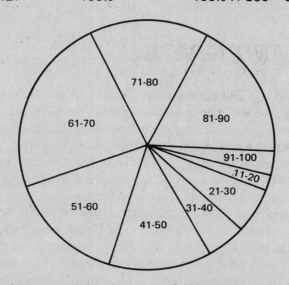

Figure 2-A. A pie chart representing the scores of 100 entering freshmen grouped into 9 class intervals.

Now remember to get yourself a protractor to measure the size of the slice of each class interval. That's that little plastic semicircle our mothers bought us in elementary school and we carried in our pencil boxes. For the circle, I recommend a compass or any perfectly round object.

Histogram

A histogram is a vertical bar graph that illustrates a frequency distribution with rectangles erected on a horizontal axis. The heights of the rectangles correspond to the frequency of the class intervals. The histogram of the academic ability scores of our 100 entering freshmen is shown in Figure 2-B. On the X-axis (horizontal), the scores or groups of scores are plotted using the class boundaries which will ensure that the vertical bars are touching. The scale of measurement on the Y-axis (vertical) is the frequency of the scores or groups of scores.

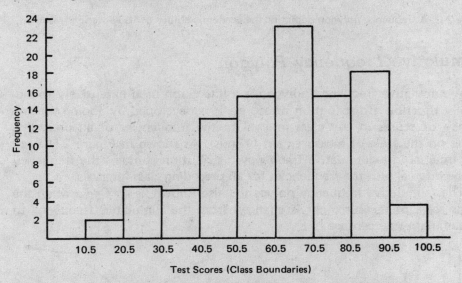

Figure 2-B: Histogram depicting academic ability of 100 entering freshmen.

Frequency Polygon

A frequency polygon is a line graph used to display a frequency distribution pictorially. As in the histogram of the academic ability of 100 freshmen (Figure 2-B), the Y-axis represents the frequency of scores or groups of scores. However, for the frequency polygon, the assumption is that the scores in each class interval are evenly distributed and can be represented on the X-axis by the class marks of each class.

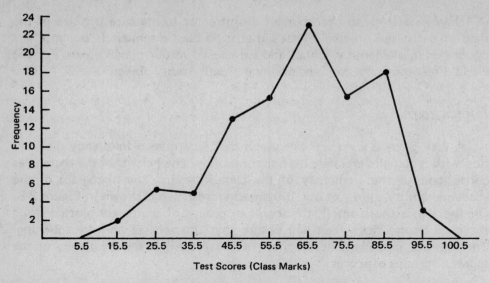

Figure 2-C. A frequency polygon depicting the academic ability of 100 entering freshmen.

Cumulative Frequency Polygon

A cumulative frequency polygon is a line graph used extensively to pictorially describe a distribution of scores. It is developed by adding the frequency of scores in any class interval to the frequencies of all preceding classes on the scale of measurement (*Y*-axis). As shown in Figure 2-D, each class interval in a cumulative frequency distribution contains the frequency for the interval plus the frequencies for all preceding class intervals.

The cumulative frequency polygon is also known as an *Ogive* when the *Y*-axis scale of measurement is changed from the cumulative frequency to the cumulative percentage.

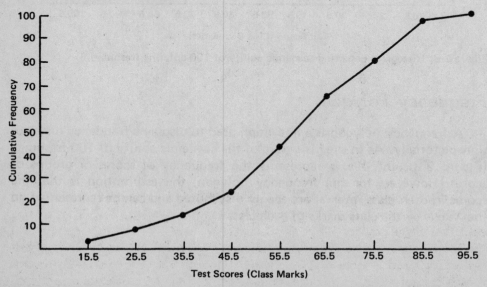

Figure 2-D. A cumulative frequency polygon depicting the academic ability of 100 entering freshmen.

Scattergram

A scattergram is a graph that illustrates the relationship between two continuous variables. For the charts and graphs we have discussed so far, the scale of measurement of the independent variable had a limited amount of discrete data points. But suppose we now want to compare the scores our 100 entering freshmen received and the number of semester hours of math taken in high school. The data for 25 of the students taken at random is shown in Table 2-E.

Table 2-E. Test Scores and the Number of Hours of High School
Math for a Sample of 25 Entering Freshmen

Student	Test Score	High School Math Hours
1	81	30
2	37	16
3	71	20
4	63	24
5	58	16
6	82	28
7	90	30
8	42	14
9	88	26
10	53	16
11	29	12
12	72	24
13	57	18
14	39	16
15	97	32
16	83	28
17	69	20
18	49	16
19	79	24
20	77	20
21	83	26
22	84	30
23	41	16
24	82	24
25	78	24

In the following scattergram (or scatter diagram or feathergram), Figure 2-E, which graphs the data in Table 2-E, dots for students 1, 2, 3, and 4 are identified in the graph.

For student 1, the dot corresponds to a score of 81 on the X variable (test scores) and a score of 30 on the Y variable (number of semester hours of high school math). The dots for students 2, 3 and 4 correspond to scores of 37, 71, and 63 and scores of 16, 20, and 24 on the *X* and *Y* measurement scales, respectively.

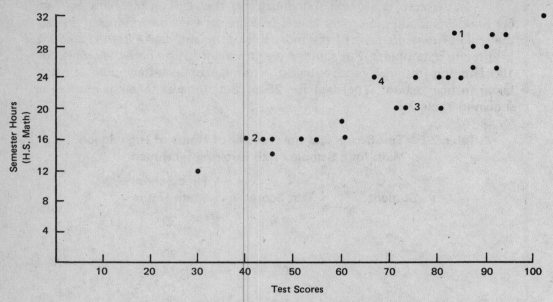

Figure 2-E. A scattergram of academic ability and the number of semester hours of high school math.

In terms of a relationship between the two variables in Figure 2-E, there is an indication that the lower scores tend to be associated with a lesser number of semester hours of high school math.

NOTE: When the data points tend to locate from lower left to upper right on the graph, the trend is generally indicative of a positive relationship between the two variables.

Now that we've gone through the various tabular and pictorial methods of summarizing and describing data, a further example of the step-by-step method might come in handy.

Table 2-F. Distribution of 40 Scores on a Spanish Quiz

65	80	82	73	65	92	71	58	84	97
59	37	40	82	71	73	80	64	70	83
42	35	58	74	63	57	81	89	90	91
60	45	42	78	81	94	82	60	87	54

Range: Highest score minus lowest score = 97 − 35 = 62.

Table 2-G. Frequency Distribution of 40 Scores

Class	Frequency	Relative Frequency	Percentage Frequency	Class Marks	Class Boundary	Cumulative Frequency
					34.5	
35-42	5	5/40 = .125	12.5%	38.5		5
					42.5	
43-50	1	1/40 = .025	2.5%	46.5		6
					50.5	
51-58	4	4/40 = .10	10.0%	54.5		10
					58.5	
59-66	7	7/40 = .175	17.5%	62.5		17
					66.5	
67-74	6	6/40 = .15	15.0%	70.5		23
					74.5	
75-82	8	8/40 = .20	20.0%	78.5		31
					82.5	
83-90	5	5/40 = .125	12.5%	86.5		36
					90.5	
91-98	4	4/40 = .10	10.0%	94.5		40
					98.5	
TOTAL:	40	40/40 = 1.00	100.0%			

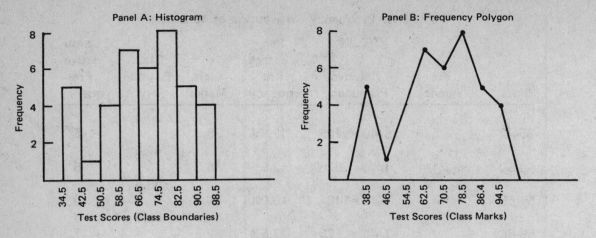

Panel A: Histogram

Panel B: Frequency Polygon

Panel C: Cumulative Frequency Polygon (0 Give)

Figure 2-F. Graphing a frequency distribution of 40 scores.

SUMMARY

Chapter 2 has introduced you to the basic procedures for the organization, classification, and summarization of numerical data in frequency distributions, charts, and graphs.

Frequency distributions tabulate the number of times a score or a group of scores occur and indicate where the scores are grouped along the scale of measurement.

Pie charts and graphs are the pictorial representations of the identical data tabulated on the frequency distribution.

In Chapter 3, we will see how the data that was collected, organized, classified, and summarized is interpreteted using quantitative measures. But first, a bit more practice in solving problems based on the material in Chapter 2 is in order.

EXERCISES

Exercise 2-4: The following is a distribution of SAT scores from a random sample of 50 high school students:

```
650  440  520  600  480  520  525  650  740  760
480  390  610  580  475  525  360  290  480  515
615  780  490  510  540  435  385  325  290  530
590  580  635  540  410  400  350  300  725  680
420  390  610  485  350  310  530  670  720  485
```

a) Construct a frequency distribution similar to Table 2-D using 10 class intervals.

b) Construct a histogram, a frequency polygon, and a pie chart to illustrate the data organized in the frequency distribution.

Exercise 2-5: Complete the following frequency distribution of Adjunct Professor Stern's teaching effectiveness which was based on responses from 120 students of statistics. Draw a pie chart of the following data:

Rating	Number of Students	Relative Frequency	Percentage Frequency	Percentage of 360 Degrees
Unsatisfactory	6			
Poor	8			
Satisfactory	18			
Good	31			
Excellent	57			
TOTAL:	120			

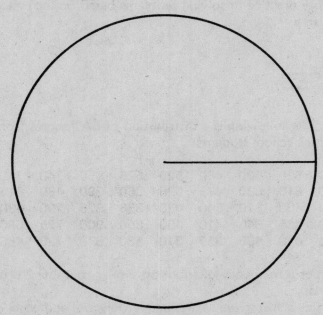

Exercise 2-6: The following is a distribution of the attendance at the non-sectarian church in Santa Monica, California, for the past year (that's 52 Sundays).

185	181	177	183
179	186	175	180
177	174	182	171
188	173	175	170
181	187	171	174
168	198	173	184
173	175	179	185
172	175	202	172
174	174	198	167
176	168	175	171
182	183	188	172
178	179	177	166
181	174	185	176

a) Using the procedures detailed in Chapter 2, develop a frequency distribution with 165-169 as the lowest class interval.
b) Construct a histogram and a frequency polygon for this distribution.

3

DESCRIPTIVE MEASUREMENTS

"The Numbers Game to Characterize Averages and Variability" or "How Do I Stack Up?"

In the preceding chapter, the collected data was put into orderly form and was then condensed into a frequency distribution table. Now, we are going to use the data to obtain and study the two most pertinent aspects of the sample data. These descriptive measurements of the data, known generally as measures of location, will help us ascertain central tendencies and dispersion of the data. The measures of central tendency are the *mean, median,* and *mode.* The measures of dispersion are the *average deviation*, the *variance*, and the *standard deviation.* The location that all of these measurements corresponds to is the center. The frequency distribution and accompanying graphs and charts in the prior chapter were certainly helpful in describing scores and groups of scores, but they did not allow for quantitative conclusions about the data or to compare it with other collected data.

To begin the introduction to the various measures of location, it is necessary to look at a form of statistical shorthand, the *Sigma notation.* The Greek letter Sigma (Σ) affords us the ability to describe the collection (addition) of a group of numbers with a single symbol. When we have n (unknown amount) observations in a sample such as x_1, x_2, x_3 all the way to x_n, we can use the Sigma notation to show the addition (sum) of the observation. To illustrate:

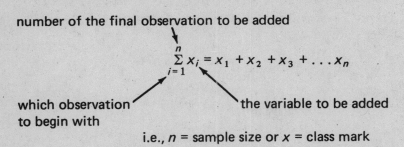

number of the final observation to be added

$$\sum_{i=1}^{n} x_i = x_1 + x_2 + x_3 + \ldots x_n$$

which observation to begin with

the variable to be added

i.e., n = sample size or x = class mark

Figure 3-A. Sigma Notation.

Examples

1) If $x_1 = 1, x_2 = 4, x_3 = -2, x_4 = 5, x_5 = 3$ and $x_6 = -4$ are 6 observations, then

$$\sum_{i=1}^{6} x_i = x_1 + x_2 + x_3 + x_4 + x_5 + x_6 = 1 + 4 + (-2) + 5 + 3 + (-4) = 7$$

and

$$\sum_{i=1}^{4} (x_i)^2 = (x_1)^2 + (x_2)^2 + (x_3)^2 + (x_4)^2 = (1)^2 + (4)^2 + (-2)^2 + (5)^2 = 1 + 16 + 4 + 25 = 46$$

2) If $x_1 = 2, x_2 = 4, x_3 = 3, f_1 = 3, f_2 = 1, f_3 = 4,$

then

$$\sum_{i=1}^{3} f_1 x_1 = (f_1)(x_1) + (f_2)(x_2) + (f_3)(x_3) = (2)(3) + (4)(1) + (3)(4) = 6 + 4 + 12 + 22$$

and

$$\sum_{i=1}^{3} f_1 x_1^2 = (f_1)(x_1)^2 + (f_2)(x_2)^2 + (f_3)(x_3)^2 = (2)(3)^2 + (4)(1)^2 + (3)(4)^2 = (2)(9) + (4)(1) + (3)(16) = 18 + 4 + 48 = 70$$

We shall be using the Sigma notation to represent the adding of additional columns in frequency distributions to aid in solving for measures of central tendency ($f \cdot x$ = frequency $\times$ class mark) and dispersion ($f \cdot x^2$ = frequency $\times$ class mark^2). Some practice problems involving the use of the Sigma notation appear at the end of this chapter.

To give you a clear picture of how the information collected and organized in a frequency distribution is employed for use in making quantitative statements, Tables 3-A and 3-B will form the foundation for the quantitative reasoning that follows.

Table 3-A. The Distribution of 138 I.Q. Tests for Freshmen Entering College on an Academic Scholarship

128	163	147	147	150	140
130	159	147	147	150	140
131	159	147	147	151	140
132	159	147	147	151	140
133	159	147	146	151	140
133	159	148	146	151	140
134	158	148	146	151	143
135	158	149	146	152	143
135	157	149	146	152	142
136	157	149	146	152	142
136	157	149	146	153	142
136	157	149	146	153	142
136	157	149	145	153	142
136	157	149	145	153	142
136	157	150	145	153	141
137	156	150	145	153	141
138	156	150	145	154	141
138	156	150	145	154	141
138	155	150	145	154	141
139	155	150	144	154	141
140	155	150	144	154	141
139	155	150	144	154	141
139	155	150	143	155	141

Table 3-B. Frequency Distribution of 138 I.Q. Scores

Class Interval	Frequency (f)	Class Mark (x)	Class Boundaries	Cumulative Frequency	f · x	fx²
			123.5			
124-128	1			1		
			128.5			
129-133	5			6		
			133.5			
134-138	13			19		
			138.5			
139-143	28			47		
			143.5			
144-148	29			76		
			148.5			
149-153	32			108		
			153.5			
154-158	24			132		
			158.5			
159-163	6			138		
			163.5			
TOTALS	138					

MEASURES OF CENTRAL TENDENCY

The mean, median, and mode (the 3 measures of the central tendency of collected data) are the quantitative measures used in the behavioral sciences to identify the location of the concentration of scores. Although commonly thought of as "averages" because all 3 are located between the highest and lowest values in the distribution, the use of the term can be confusing because the 3 measurements represent numerical values that can be quite diverse.

Since the purpose of this book is to remove, not add to the confusion surrounding the study of statistics, we will describe the measures of central tendency and, later, the measures of dispersion.

Mean

The mean is the most frequently used measure of central tendency. Denoted by the symbol $\overline{x}$, the mean is defined as the arithmetic average of all the scores or groups of scores in a distribution.

For Ungrouped Data

$$\text{Mean } (\overline{x}) = \frac{\text{sum of the observations}}{\text{number of observations}} = \frac{\Sigma x}{n}$$

where,

x = the scores in a distribution
n = the number of observations in the distribution

The idea is to simply add up all the values and then divide by the number of observations.

Examples

1) Nine students received the following scores on a surprise quiz: 67, 53, 81, 69, 77, 91, 66, 76, 50. Find the mean.

$$\text{Mean} = \frac{\Sigma x}{n} = \frac{67 + 53 + 81 + 69 + 77 + 91 + 66 + 76 + 50}{9} = \frac{630}{9}$$

$$\overline{x} = 70$$

2) Find the mean of the following SAT scores.

593	512	480	485
673	614	502	560
573	704	485	580
716	626	610	670

$$\text{Mean } (\overline{x}) = \frac{\Sigma x}{n} = \frac{9383}{16} = 586.4375$$

For Grouped Data

$$\text{Mean } (\overline{x}) = \frac{\text{frequency of each class} \times \text{the score or class mark}}{\text{total number of observations}}$$

$$\overline{x} = \frac{\Sigma f \cdot x}{n} \text{ where,}$$

f = frequency
x = score or class mark
n = number of observations in the sample

Instead of adding the individual scores in a distribution, it is necessary to add a column to the frequency distribution. The total of this column ($f \cdot x$) will become the numerator of the equation, $\overline{x} = \frac{\Sigma f \cdot x}{n}$.

Score (in inches) (x)	Frequency (f)	f · x
56	6	336
57	8	456
58	14	812
59	11	649
60	5	300
61	6	366
TOTALS	50	2919

Figure 3-B. The heights of 50 fifth grade girls.

Using the data in Figure 3-B, the mean (x) = the sum of the column, f · x, divided by the total number of observations.

$$\overline{x} = \frac{\Sigma f \cdot x}{n} = \frac{2919}{50} = 58.38 \text{ inches}$$

Class	Frequency (f)	Class Mark (x)	f · x
43-48	4	45.5	182.0
49-54	1	51.5	51.5
55-60	5	57.5	287.5
61-66	3	63.5	190.5
67-72	9	69.5	625.5
73-78	12	75.5	906.0
79-84	9	81.5	733.5
85-90	7	87.5	612.5
TOTALS	50		3589.0

Figure 3-C. Test scores of 50 students.

To solve for the mean when the collected data is grouped into classes, the class mark (midpoint) of each class represents the "x" in our formula. Based on the data in Figure 3-C, the solution is

$$\text{Mean } (\overline{x}) = \frac{\Sigma f \cdot x}{n} = \frac{3589.0}{50} = 71.78$$

It is now time to return to Table 3-B and solve for the mean score for the 138 students on the I.Q. test.

Use one column, number six, approximately headed "f · x" and using the total from that column in the formula, solve for the mean (x̄).

Median

When data is placed in array (high-to-low, or low-to-high), the median represents the point in the data where one-half (50%) of the scores fall below that point and one-half (50%) fall above it. The median income for a population means that 50% of the families earn less than that amount and 50% of the families earn more.

For Ungrouped Data

When there's an odd number of observations placed in array, the median corresponds to the $(\frac{n+1}{2})^{th}$ largest observation.

Examples

1) There are 7 people in a room, ages 17, 19, 20, 27, 29, 36, and 43. The median corresponds to the $(\frac{n+1}{2})^{th}$ observation.
    i)   Since $n = 7$, $\frac{n+1}{2} = \frac{7+1}{2} = \frac{8}{2} = 4$
    ii) The 4th observation, the median, is 27. There are then 3 people younger and 3 people older.

2) If there are 103 observations in array, the median will correspond to the $(\frac{n+1}{2})^{th}$ largest observation.
    i) $\frac{103+1}{2} = \frac{104}{2} = 52$. The 52nd largest observation will correspond to the median.
    ii) When there is an even number of observations in array, the median corresponds to the midpoint between the $(\frac{n}{2})^{th}$ and the $(\frac{n}{2} + 1)^{th}$ largest observations.

Examples

1) There are 8 people in a room, ages 17, 19, 20, 21, 29, 36, 43, and 50. The median corresponds to the midpoint between the $(\frac{n}{2})^{th}$ and $(\frac{n}{2} + 1)^{th}$ observations. $\frac{n}{2} = \frac{8}{2} = 4$. $\frac{n}{2} + 1 = \frac{8}{2} + 1 = 4 + 1 = 5$. The median is therefore the midpoint between the 4th and 5th observation. The midpoint between 21 and 29 is 25. With a median of 25, there will be 4 people younger and 4 people older. And with an even number of observations, the median of 25 does not represent one of the ages of the people present.

2) If there are 100 observations in array, the median corresponds to the midpoint of the $\frac{n}{2}$ (50th) and the $\frac{n}{2} + 1$ (51st) observations. That will leave 50 below the median and 50 above.

For Grouped Data

As you've just discovered, the median is different from odd and even numbers of data when the data is not grouped. However, if the data is grouped into classes, the median is simply defined as the $(\frac{n}{2})^{th}$ largest observation. Since the observations, though in array, are grouped in classes, the median will fall into one of the classes. Although the process is somewhat more complicated, there is a general formula for finding the median of grouped data.

$$\text{median} = L + [\tfrac{c}{f}(\tfrac{n}{2} - F)]$$

where

L = the lower class boundary (lower in value) of the class containing the median.

n = the total number of observations.

F = the cumulative frequency up to the lower limit of the median class (the total of all preceding classes).

f = the frequency of the class containing the median.

c = the length of the class interval containing the median.

In the following examples, we will go through the formula step by step.

Examples

1) The following table gives the distribution of scores on the ages of people on a bus.

Class	Frequency	Class Boundaries	Cumulative Frequency
		9.5	
10-19	16		16
		19.5	
20-29	18		34
		29.5	
30-39	14		48
		39.5	
40-49	15		63
		49.5	
50-59	17		80
		59.5	
60-69	10		90
		69.5	
TOTALS	90		

i) Since there are 90 observations in the sample, the median will correspond to the $\tfrac{n}{2}$th or the $\tfrac{90}{2}$th observation; that is, the 45th largest observation. The first and key step in determining the median is to determine in which class the 45th observation falls. The first 2 classes have a cumulative frequency of 34 observations. We need another 11 observations to reach 45. Therefore, the 45th observation must fall in the next class which contains 14 observations. The median class, the basis for solving our formula, is 30-39. The rest is easy.

L = lower boundary of median class. For our class 30-39, the lower boundary is 29.5.

n = total frequency = 90.

F = total frequency up to lower limit of median class. That is the frequency of classes 1 and 2 which are 16 + 18 = 34.

f = frequency of the median class = 14.

c = length of median class = 10.

ii) median = $L + [\frac{c}{f}(\frac{n}{2} - F)]$

iii) median = $29.5 + [\frac{10}{14}(\frac{90}{2} - 34)]$

$= 29.5 + [\frac{10}{14}(45 - 34)] = 29.5 + [\frac{10}{14}(11)]$

$= 29.5 + [7.857]$

$= 37.357$

This means that 50% or 45 of our 90 scores will fall below 37.357 and 50% or 45 will fall above.

2)

Class	Frequency	Class Boundaries	Cumulative Frequency
		.5	
1-25	11		11
		25.5	
26-50	13		24
		50.5	
51-75	16		40
		75.5	
76-100	10		50
		100.5	
TOTALS	50		

i) Because there are 50 observations, the median corresponds to the $\frac{n}{2}$th or 25th observation. The first 2 classes contain 24 observations. The 25th will fall in the next class 51-75.

ii) Median = $L + [\frac{c}{f}(\frac{n}{2} - F)]$

$= 50.5 + [\frac{25}{16}(\frac{50}{2} - 24)]$

$= 50.5 + [\frac{25}{16}(25 - 24)]$

$= 50.5 + [\frac{25}{16}(1)]$

$= 50.5 + 1.56$

$= 52.06$

It's again time to return to Table 3-B and solve for the median. Keep the formula and the explanation handy. It's not necessary to memorize it. What we are striving for is understanding.

Mode

This is the simplest of the 3 measures of central tendency to compute. The mode is the score (observation) or group of scores that occurs most frequently. If 2 scores or groups of scores are tied for the highest frequency, we say the data is bimodal. When the data is divided into classes of equal size, the class with the highest frequency is called the *modal* class. The class mark of the modal class is then defined as the mode. If the class sizes are different, then it is necessary to determine the modal class by the density.

$$\text{Frequency density} = \frac{\text{frequency}}{\text{class size}}$$

The class with the highest density is thus the modal class.

Examples

1) Using the data in Examples 2) on page 28, the modal class, the one with the highest frequency, was 51-75. The mode, the class mark, would be $\frac{51 + 75}{2} = 63$.

   NOTE: The class sizes were all the same.

2) Using the ages of 12 people in a room aged 19, 21, 21, 24, 26, 26, 27, 29, 31, 35, 37 and 39, we see that both 21 and 26 have a frequency of 2. The data is therefore bimodal. Now, return to Table 3-B and solve for the mode. After that, try the exercises that follow.

EXERCISES

Exercise 3-1: Find the mean, median, and mode for the following sets of observations:

   a) 2, 6, 9, 9, 11, 11, 11, 14, 17
   b) 16, 18, 19, 19, 20, 24, 27, 27, 27, 29

Exercise 3-2.

Class	Frequency	Class Mark	Class Boundaries	Cumulative Frequency
44-53	9			
54-63	10			
64-73	10			
74-83	11			
84-93	11			
94-100	9			
TOTALS	60			

Determine the mean, median, and mode for the preceding distribution of 60 test scores.

Comparison of the Mean, Median, and Mode

Which is the most accurate measure of the central tendency of a set of data? Selection of the most appropriate quantitative measurement is dependent on the variable being measured. The use to be made of the collected data should influence the choice of a specific measure. If the variable is qualitative, such as an evaluation of a president or teacher, the mode is usually most appropriate. With quantitative variables that have small samples with extreme scores, the median is preferable. When the sample is large and the variable is quantitative, the mean has a distinct advantage. Both the mode and median are usually unaffected by any extreme scores. The median, which is useful with quantitative variables, is worth little when used to describe qualitative variables. The following chart illustrates the benefits and problems inherent in the 3 distinct measurements.

Doctor (6)	$120,000	
Dentist (7)	60,000	
Lawyer (7)	45,000	$\bar{x}$ = $44,268
Policeman (1)	27,000	median = $27,000
Salesman (5)	24,000	mode = $11,000
Civil Service (6)	19,000	
Writer (9)	11,000	

Figure 3-D. The distribution of average salaries for 41 members of a racquetball club.

Three quantitative measurements; 3 diverse conclusions. You judge for yourself which is most valid. The diversion makes us aware of the misuse, as well as the use, of statistics.

MEASURES OF DISPERSION

As explained earlier in this chapter, the measures of central tendency describe only one of the important parameters (characteristics) of a frequency distribution. For a better understanding of the data, it is important to know how the data is spread throughout the distribution. The spread is also known as the scatter, the variability, and the dispersion of the data. The 3 measures to be considered here are the average (mean) deviation, the variance, and the standard deviation. These measures will aid us in making comparisons, because quantitative measurements are preferable to pictorial comparisons in scientific research, and data with extreme differences in the spread will still give us similar, if not exact, measures of central tendency.

Examples

1)   45, 50, 55, 60, 65
     $\bar{x} = 55$ median = 55
2)   5, 10, 55, 100, 105
     $\bar{x} = 55$ median = 55

Both of the examples had identical measures of central tendency even though the spread of the data is dissimilar. In 1) the range of the data = 65 − 45 = 20. In 2), the range = 105 − 5 = 100. Thus, the data in 2) is definitely more variable and thus the need for additional measures of location to help in defining the collected data. The measures of dispersion are all based on the distance (deviation) between the scores and the mean. The difference between the mean and the score or class mark for a group of scores gives us the deviation.

Average (Mean) Deviation

This measure of dispersion or variability is defined as the absolute difference (deviation) between the observations in a sample and the mean divided by the total number of observations in the sample. The word "absolute," which is represented by the sign "| |" in mathematics, simply means that we take the value of a number without regards to + or −.

For Ungrouped Data

$$\text{Average (mean) deviation} = \frac{\sum_{i=1}^{n} |x - \bar{x}|}{n}$$

Examples

1) We have a sample of observations that consists of 45, 50, 55, 60, 65. The mean is 55.

Average deviation

$$= \frac{|45 - 55| + |50 - 55| + |55 - 55| + |60 - 55| + |65 - 55|}{5}$$

$$= \frac{|-10| + |-5| + |0| + |5| + |10|}{5}$$

$$= \frac{|10| + |5| + |0| + |5| + |10|}{5}$$

$$= \frac{|30|}{5} = \frac{30}{5} = 6$$

Thus, on the average, each observation is 6 units from the mean.

2) We have a second sample consisting of the observations 5, 10, 55, 100, 105.

Average deviation

$$= \frac{|5 - 55| + |10 - 55| + |55 - 55| + |100 - 55| + |105 - 55|}{5}$$

$$= \frac{|-50| + |-45| + |0| + |45| + |50|}{5}$$

$$= \frac{|50| + |45| + |0| + |45| + |50|}{5}$$

$$= \frac{|190|}{5} = \frac{190}{5} = 38$$

Thus, in our second sample with the same mean the average deviation is 38 units from the mean.

For Grouped Data

When the collected data is grouped into classes, it is necessary to take the difference between the class marks and the mean. The deviation is then multiplied by the frequency and the total is divided by the number of observations in the distribution.

$$\text{Average deviation} = \frac{\sum_{i=1}^{n} f|x - \bar{x}|}{n}$$

Examples

1) Based on the following distribution of the weights (in ounces) of cartons of ice cream, we will find the average deviation. The first step is to find the mean ($\bar{x}$).

Class	f	Class Mark (x)	f · x	|x − 17.11|	f|x − 17.11|
14.0-14.9	11	14.45	158.95	2.66	11 X 2.66 = 29.26
15.0-15.9	13	15.45	200.85	1.66	21.58
16.0-16.9	21	16.45	345.45	.66	13.86
17.0-17.9	15	17.45	261.75	.34	5.10
18.0-18.9	11	18.45	202.95	1.34	14.74
19.0-19.9	19	19.45	369.55	2.34	44.46
TOTALS	90		1539.50		129.00

i) $\quad$ Mean $= \Sigma \dfrac{f \cdot x}{n} = \dfrac{1539.50}{90} = 17.11$

ii) Average deviation $= \Sigma \dfrac{f \cdot |x - x|}{n} = \dfrac{129.00}{90} = 1.43$

Class	f	x	f · x	|x − 21.23|	f|x − 21.23|
10-13	3	11.5	34.5	9.73	3 X 9.73 = 29.19
14-17	4	15.5	62.0	5.73	22.92
18-21	7	19.5	136.5	1.73	12.11
22-25	9	23.5	211.5	2.27	20.43
26-29	7	27.5	192.5	6.27	43.89
TOTALS	30		637.0		128.54

i) $\quad \bar{x} = \dfrac{\Sigma f \cdot x}{n} = \dfrac{637}{30} = 21.23$

ii) Average deviation $= \dfrac{\Sigma f \cdot |x - \bar{x}|}{n} = \dfrac{128.54}{30} = 4.28$

The average (mean) deviation is the easiest measure of dispersion to calculate, the simplest to understand and use, and it uses every observation in the sample. But the average deviation lacks a strong mathematical relationship with the specific scores in the distribution. What the average deviation can tell us is that the distributions with the larger mean deviations have the greater spreads on both sides of the mean. For practice, refer to Table 3-B and solve for the average deviation.

Variance

To avoid any difficulty associated with using absolute values for the deviations from the mean, square the deviations. Absolute values disregard positive (+) and negative (−) values. This works, but the mathematically correct way to avoid negative (−) deviations is to square the negative deviations. With a score of 50 and a mean of 55, the deviation is −5. If we square the deviation (−5) × (−5), we end up with (+)25. Later, if we take the square root of 25 ($\sqrt{25}$), the result, which is +5, has removed the negative sign. What we obtain using this method by definition is the *variance*, the sum of the squared deviations about the mean. When we are discussing the variance of a population, we use the symbol σ^2. For samples, use s^2.

For Ungrouped Data

$$\text{Variance} = \frac{\Sigma(x - \bar{x})^2}{n}$$

where,

$\bar{x}$ = mean of the distribution
n = number of observations in the data

Examples

1) We have a distribution of 10 sixth grade students whose weight in pounds was: 90, 95, 96, 97, 98, 101, 103, 105, 106, and 109.

   i) Mean ($\bar{x}$) $= \dfrac{1000}{10} = 100$ pounds

   ii) Variance (σ^2) $=$

   $$\frac{\begin{array}{c}(90 - 100)^2 + (95 - 100)^2 + (96 - 100)^2 + \\ (97 - 100)^2 + (98 - 100)^2 + (101 - 100)^2 + \\ (103 - 100)^2 + (105 - 100)^2 + (106 - 100)^2 + \\ (109 - 100)^2\end{array}}{10}$$

   $$= \frac{\begin{array}{c}(-10)^2 + (-5)^2 + (-4)^2 + (-3)^2 + (-2)^2 + \\ (1)^2 + (3)^2 + (5)^2 + (6)^2 + (9)^2\end{array}}{10}$$

   $$= \frac{100 + 25 + 16 + 9 + 4 + 1 + 9 + 25 + 36 + 81}{10}$$

   Variance $= \dfrac{306}{10} = 30.6$ pounds squared

2) The number of minutes a lawyer spent making 6 phone calls was: 3, 8, 9, 11, 15, and 20 minutes.

i) Mean ($\bar{x}$) $= \dfrac{66}{6} = 11$ minutes

ii) Variance (σ^2) $= (3-11)^2 + (8-11)^2 + (9-11)^2 +$
$$\dfrac{(11-11)^2 + (15-11)^2 + (20-11)^2}{n}$$

$$= \dfrac{(-8)^2 + (-3)^2 + (-2)^2 + (0)^2 + (4)^2 + (9)^2}{6}$$

$$= \dfrac{64 + 9 + 4 + 0 + 16 + 81}{6}$$

$$= \dfrac{174}{6} = 29 \text{ minutes squared}$$

For Grouped Data

When the data is grouped into classes, we use the class mark as "x" in our formula. The square of the deviations is then multiplied by the frequency of each class. The total is then divided by the number of observations in the distribution.

$$\text{Variance} = \dfrac{f(x - \bar{x})^2}{n}$$

NOTE!! Although the preceding formula is mathematically sound and correct, it requires two columns of calculation in our frequency distributions. The purpose of this book is to make the job of learning statistics less painful, and to that lofty aim the following formula will be substituted when solving for the variance.

$$\text{Variance} = \Sigma \dfrac{f \cdot x^2}{n} - (\bar{x})^2$$

This formula allows us to arrive at the right answer by only subtracting the square of mean *one* time.

Examples

1) The heights of 40 children are shown below:

(x) Height in Inches	f	f · x	f · x²	fx · x
59	8	472	8(59)² or 472(59) =	27848
60	5	300		18000
61	12	732		44652
62	9	558		34596
63	6	378		23814
TOTALS	40	2440		148910

i) Mean $(\overline{x}) = \dfrac{2440}{40} = 61$

ii) Variance $= \Sigma \dfrac{f \cdot x^2}{n} - (\overline{x})^2$

$= \dfrac{148{,}910}{40} - (61)^2$

$= 3722.75 - 3721$

$= 1.75$ inches squared

NOTE: The smaller the variance, the smaller the dispersion of the data about the mean.

Class	f	Class Mark (x)	f · x	f · x²
56-60	4	58	232	13456
61-65	3	63	189	11907
66-70	3	68	204	13872
71-75	4	73	292	21316
76-80	5	78	390	30420
81-85	2	83	166	13778
86-90	4	88	352	30976
91-95	3	93	279	25947
96-100	2	98	196	19208
TOTALS	30		2300	180880

i) $\quad \bar{x} = \Sigma \dfrac{f \cdot x}{n} = \dfrac{2300}{30} = 76.67$

ii) $\quad$ Variance $= \Sigma \dfrac{f \cdot x^2}{n} - (\bar{x})^2 = \dfrac{180880}{30} - (76.67)^2$

$\qquad\qquad\qquad = 6029.33 - 5878.29$

$\qquad\qquad\qquad = 151.04$ units squared

Standard Deviation

Simply defined as the square root of the variance, the standard deviation is the most useful measure of dispersion.

For Ungrouped Data

Standard deviation $(\sigma) = \sqrt{\Sigma \dfrac{(x - \bar{x})^2}{n}}$

For Grouped Data

1) Standard deviation $(\sigma) = \sqrt{\Sigma \dfrac{f(x - \bar{x})^2}{n}}$

2)* Standard deviation $(\sigma) = \sqrt{\Sigma \dfrac{fx^2}{n} - (\bar{x})^2}$

*Use this formula to solve problems.

The variance is a deviation measured in square units. The standard deviation, the square root of the variance, is the measure of dispersion in the original unit of measurement. And because the standard deviation is in the original unit of measurement, it has certain advantages. For example, the standard deviation (standard margin of error) can be used in quality control to insure products are sold according to prescribed specifications. In intelligence or aptitude tests, the standard deviation allows us to compare student's results based on an established mean score.

For the last time, I want you now to turn to Table 3-B and solve for the variance and the standard deviation. (If you're using a calculator, solve for the variance and then just push the square root button to obtain the standard deviation.)

SUMMARY

Chapter 3 has covered the procedures for computing the quantitative measures of central tendency and dispersion. The 3 measures of central tendency—mean, median, and mode—quantitatively define the central location of a value between the extreme limits in a distribution. This allows us to compare scores to a central value, such as the median or mean income of families in California. You can then see how you stack up in comparison. The 3 measures of dispersion—average deviation, variance, and standard deviation—quantitatively measure the extent to which scores in a distribution are spread throughout.

Together, the measures of location enable us to accurately describe a distribution and to adequately understand the collected data in order to make practical quantitative conclusions. Data might as well be in a vacuum if it is not interpreted for use in research.

EXERCISES

Exercise 3-3: The following are the number of parents attending 25 biweekly PTA meetings.

120	100	120	110	60
150	140	170	90	120
130	80	70	150	140
150	180	190	140	100
140	140	160	80	90

a) Solve for the mean, median, and mode.
b) Solve for the average deviation, variance, and standard deviation.

Exercise 3-4: The following distribution represents the scores of 80 students on a math test.

Class (Scores)	Frequency (f)
30-39	7
40-49	9
50-59	14
60-69	13
70-79	10
80-89	17
90-99	10
TOTALS	80

a) Solve for the mean.
b) Solve for the variance and standard deviation.

We will be leaving the areas of descriptive statistics and quantitative measurements to move into the world of probability which forms the basis of the next 3 chapters. The laws of probability built Las Vegas and Monte Carlo. Learn them and reduce the odds. Learn them and beat your friends at backgammon. Use them in business and avoid the temptation to make emotional decisions.

4

PROBABILITY

"The Laws of Chance"

The empirical world is replete with random events that affect our daily lives and destinies. The notion of probability, even in a nonmathematical sense, is familiar to everyone. Who hasn't wondered if Sunday's outdoor party would be spared inclement weather? What newspaper editor has not pondered the probability of nuclear war? What sports fan has not argued the probability of his favorite team winning the championship? What mother has not mused on the probability of her only daughter marrying a bum?

Good questions, but there is no exact (in mathematical terms) way to determine the true probability of their occurrence. Although most of us draw on the concept of probability in everyday discussions and disagreements, there is generally a large degree of imprecision involved with statements concerning probability.

Probability deals with random events. A random event by definition is an event (experiment) in which each of the possible outcomes has an equal chance of occurring. For example, when a coin is flipped, there are only 2 outcomes—heads or tails—and unless the coin is weighted to favor one side, there is an equal (50%) chance of heads or tails. With a single die (that's one-half of a pair of dice) that has six numbers, 1 through 6, each has an equal chance of occurring when the die is rolled. When selecting one card from a deck of 52, there is an equal likelihood of selecting any card.

PROBABILITY OF SIMPLE EVENTS

In a mathematical sense, probability is defined as the relative frequency of an event or experiment. Probability is the ratio of the number of favorable outcomes to the total number of possible outcomes.

$$\text{Probability (Event A)} = \frac{\text{number of favorable outcomes}}{\text{total number of outcomes}}$$

where the total number of outcomes = those favoring A + those *not* favoring A.

NOTE: It is *essential* to understand that the total probability of any event or experiment equals 1. The probability of an event can be 0, if there is no probability of occurrence. This is called an impossible event and it is based on our empirical world. One example of an impossible event is the probability of a man having a baby. The probability of an event can be 1. This is a simple event with only one outcome. An example is the probability of the sun rising tomorrow. For the purposes of understanding probability theory, the results of an experiment must fall between 0 and 1, inclusive.

$$0 \leq \text{Probability of Event A} \leq 1$$

All of the probabilities in any experiment must total 1.

EXAMPLES

1) When flipping a coin there are two possible outcomes.

   i) Probability (head) = $\dfrac{\text{number of favorable outcomes}}{\text{total number of outcomes}}$ = 1/2

   ii) Probability (tail) = $\dfrac{\text{number of favorable outcomes}}{\text{total number of outcomes}}$ = 1/2

   iii) Probability of head or tail = 1/2 + 1/2 = 1

2) When rolling 1 die there are 6 possible outcomes. Each has a probability of 1/6. When adding all the probabilities of the possible outcomes, we get

$$1/6 + 1/6 + 1/6 + 1/6 + 1/6 + 1/6 = 6/6 = 1.$$

NOTE: Since the total probability of an event equals 1 and this includes both favorable and nonfavorable outcomes, the probability of Event A + the probability of *not* Event A equals 1.

Probability (A) + probability (*not* A) = 1, therefore
probability (*not* A) = 1 − probability (A)

Example: Flipping a coin

a)   The probability (heads) = 1/2
b)   The probability of (tails or *not* heads) = 1/2
c)   The probability (*not* heads) = 1 −
     probability (*not* heads) = 1 − 1/2 = 1/2

THE PROBABILITY RULES FOR COMPOUND EVENTS

In contrast to a single event, such as flipping a coin or picking a card, there are experiments that involve two or more single events. A compound event involves the probability of the joint occurrence of two (or more) events in the "either-or" or "both" options. "Either-or" consists of adding the probabilities, whereas "both" consists of multiplying the probabilities. Formulas to solve the distinct problems are available and easy to use.

```
┌─────────────────────────────┐
│ A  B  C  D  E  F  G  H  I  J │
│ K  L  M  N  O  P  Q  R  S  T │
└─────────────────────────────┘
```

Figure 4-A. The sample space for the first 20 letters of the alphabet.

Rules for Addition of Compound Events

For Mutually Exclusive Events

By definition, mutually exclusive events have no outcomes in common. The probability of selecting a vowel or the letter "S" in Figure 4-A involves two events. Since it is impossible to obtain both a vowel and an "S" from one random selection, the formula used is:

Probability (A or B) = probability (A) + probability (B)
when probability (A and B) = 0

Based on Figure 4-A, the probability of the union of a vowel or an "S" is:

$$P \text{ (vowel or S)} = P \text{ (vowel)} + P(S)$$

$$P \text{ (vowel)} = \frac{\text{number of favorable outcomes}}{\text{total number of outcomes}} = 4/20$$

$$P(S) = \frac{\text{number of favorable outcomes}}{\text{total number of outcomes}} = 1/20$$

$$P \text{ (vowel or S)} = 4/20 + 1/20 = 5/20 = .25$$

Examples

1) What is the probability that on a single roll of the die, the number is either even or a 3? Since it's impossible on one roll to get an even number and a 3, we get:
    i) P (even number or 3) = P (even number) + P(3)
    ii) P (even number) = 3/6, P(3) = 1/6
    iii) P (even number or 3) = 3/6 + 1/6 = 4/6

2) Out of a sample of 10,000 men, the probability that a man picked at random weighs over 190 pounds is .25. The probability that a man weighs less than 135 pounds is .15. What is the probability that a man picked at random weighs either over 190 pounds or under 135 pounds? Since it is impossible for 1 man to be both over 190 and under 135 pounds, we have:

$$P \text{ (over 190 or under 135)} = P \text{ (over 190)} + P \text{ (under 135)}$$
$$= .25 + .15$$
$$= .40$$

3) What is the probability that a man picked at random weighs between 135 and 190 pounds? Between 135 and 190 pounds means not less than 135 and not more than 190. Now we have:

   i) P *not* (over 190 or under 135) = 1 — (probability of over 190 or under 135)

$$= 1 - .40 = .60$$

   ii) Did you remember that the total probability (favorable and not favorable) must always equal 1?

A	B	C	D	E	F	G	H	I	J
K	L	M	N	O	P	Q	R	S	T
		Event A					Event B		

Figure 4-B. The sample spaces for Events A and B for the first 20 letters of the alphabet.

For Non-Mutually Exclusive Events

In more complex probability experiments, involving non-mutually exclusive events, the joint occurrence, the probability (A and B), is possible. For these experiments the requisite formula is:

$$\text{Probability (A or B)} = P(A) + P(B) - P(A \text{ and } B), \text{ when}$$
$$P(A \text{ and } B) \neq 0$$

In Figure 4-B, the letter "O" is coincident in both Event A and Event B. In order to avoid the letter "O" being counted twice in the union of A and B, it becomes necessary to subtract any part of a sample space that coexists in two events. In our example, P(A and B) applies to the probability of the letter "O".

P (A or B) = P ("O" in Event A) + P ("O" in Event B) — P ("O" in both Event A and Event B)
$$= 1/10 + 1/6 - 1/15 = 6/60 + 10/60 - 4/60 = 12/60 = .20$$

The probability (A and B) refers to the probability that the sample events will show up in the two compound events.

Examples

1) What is the probability of selecting an ace or a spade from a deck of 52 cards?

   i)   P (A or B) = P (ace) + P (spade) — P (ace and spade)
$$= 4/52 + 13/52 - 1/52$$
$$= 16/52$$

   ii) Because the ace of spades was counted twice—first as an ace and then as a spade—it was necessary to subtract it once for the concurrence. The 16 favorable outcomes in the answer then represents the 13 spades plus the aces of clubs, diamonds, and hearts.

2) Debra is taking two college entrance exams, in English and in Math. The probability that she will pass the English exam is .75. The probability that she will fail the Math exam is .20. The probability that she will pass both exams is .65. What is the probability that Debra will pass either the Math or the English exam?

   i) The two events are not mutually exclusive because Debra can pass both events. Therefore, the formula is:

$$P \text{ (A or B)} = P(A) + P(B) - P(A \text{ and } B)$$
$$P \text{ (English or Math)} = P \text{ (English)} + P \text{ (Math)} - P \text{ (Math and English)}$$
$$P \text{ (English)} = .75$$
$$P \text{ (not Math)} = .20$$
$$P \text{ (Math)} = 1 - .20 = .80 \text{ because the probabilities of Math}$$
$$\text{and not Math must equal 1.}$$

   ii) $P \text{ (English or Math)} = .75 + .8 - .65$
$$= 1.55 - .65 = .90$$

There is a 90% probability Debra will pass either the English or the Math exam.

Rules for Multiplication of Compound Events

For Statistically Independent Events

In the previous sections, we have discussed the probability of the occurrence of a simple (single or compound) event. In the following sections, you will learn to calculate the probabilities of a combination (2 or more) of events. When the occurrence or nonoccurrence of Event A has no effect on the occurrence of Event B, we say the two events, A and B, are statistically independent. If events are determined to be independent, then the probability of both occurring is derived by multiplying the probabilities of the individual events. This is true for 2 or more independent events, such as the flipping of a coin. Each time the coin is flipped, the probability of a head or tail is independent of the any other flips. Ergo, the formula for the multiplication of independent events is:

$$P(A \text{ and } B) = P(A) \cdot P(B)$$
$$P(A \text{ and } B \text{ and } C) = P(A) \cdot P(B) \cdot P(C)$$

When flipping coins we get,

a) one coin — $P \text{ (head)} = 1/2$
b) 2 coins — $P \text{ (head and head)} = (1/2)(1/2) = 1/4$. The probabilities of two tails is also 1/4. The probabilities of a head and a tail are 1/4 and the probability of a tail and a head is 1/4. The 4 probabilities together $(1/4 + 1/4 + 1/4 + 1/4)$ equal 1, the total probability.
c) 3 coins — $P \text{ (head and head and head)} = (1/2)(1/2)(1/2) = 1/8$.

When a coin is flipped 3 times, out of the 8 total outcomes, one will be 3 heads.

Referring to Figure 4-A, we can calculate the probability of selecting a letter at random, replacing it in the sample, and on a second random selection choosing the same letter.

$$P(N \text{ and } N) = P(N) \cdot P(N) = 1/20 \cdot 1/20 = 1/400$$

Example

1) If 2 dice are tossed, what is the probability of obtaining a 4 on the first die and an odd number on the second?

    i)   P(4 and an odd number) = P(4) · P (odd number)

$$= 1/6 \cdot 3/6 = 3/36 \text{ or } 1/12.$$

NOTE: When tossing 2 dice, the number of total outcomes is also multiplied. Therefore there are 6 × 6 or 36 possible outcomes when 2 dice are tossed.

Table 4-A. The Results of Tossing Two Dice

Results	Number of Favorable Outcomes	Probability
2	1	1/36 = 2.8%
3	2	2/36 = 5.5%
4	3	3/36 = 8.3%
5	4	4/36 = 11.1%
6	5	5/36 = 13.9%
7	6	6/36 = 16.7%
8	5	5/36 = 13.9%
9	4	4/36 = 11.1%
10	3	3/36 = 8.3%
11	2	2/36 = 5.5%
12	1	1/36 = 2.8%
	36	36/36 = 99.9% (close enough)

    ii)   When 2 dice are tossed, the smallest result is 2 (1 and 1) and the largest is 12 (6 and 6). (I hope this data will benefit you financially on your next visit to a Vegas craps table.)

2) What is the probability of drawing a face card (Jack, Queen, King) on your first selection from a deck of 52 cards, and without replacement, drawing an ace on the second selection?

$$P \text{ (face card and ace)} = P \text{ (face card)} \times P \text{ (ace)}$$
$$= 12/52 \times 4/51 = 48/2652$$

For Non-Independent (Dependent) Events

When the outcome of one event is dependent on the occurrence or non-occurrence of another event, the formula is:

$$P(A \text{ and } B) = P(A) \cdot P(A/B)$$
$$\text{or}$$
$$P(A \text{ and } B) = P(B) \cdot P(B/A)$$

This is *conditional probability* and I think it is easier to understand if the formula is adjusted algebraically, giving us:

$$P(A/B) = \frac{P(A \text{ and } B)}{P(B)}$$

a) P(A/B) means the probability of Event A given that Event B has occurred

b) P(B) is the probability of the given (conditional) event

NOTE: In conditional or dependent probability problems, the condition changes the total number of outcomes in the probability fraction by placing a limitation on the number.

Examples

1) Calculating the probability of a simple event, the probability of selecting an even number from the sample 1, 2, 3, 4, 5, 6, 7, 8, 9, 10 is 5/10 or 1/2. Given that Event A has occurred, what is the probability of selecting a number greater than 7?

   i) Using the formula for conditional probability we get:

$$P(B/A) = \frac{P(A \text{ and } B)}{P(A)}$$

   ii) $P\left(\dfrac{\text{number greater}}{\text{than 7}} \Big/ \dfrac{\text{number is}}{\text{even}}\right) = \dfrac{P(\text{greater than 7 and even})}{P(\text{even})}$

$$= \frac{2/10(\text{numbers 8 and 10})}{5/10}$$

$$= 2/10 \div 5/10 = 2/10 \times 10/5 = 2/5$$

   iii) Event A limited the number of total outcomes from 10 to 5. Only the even numbers are left. Out of those 5—2, 3, 6, 8, and 10—the numbers 8 and 10 are greater than 7.

   iv) Using the original formula, we would have calculated for the probability of Event A and Event B.

$$P(A \text{ and } B) = P(B) \cdot P(B/A)$$
$$= 5/10 \cdot 2/5$$
$$= 10/50 = 2/10$$

2) Bill attempts to kick two field goals. The probability that the first kick is successful is .70. The probability that Bill is successful on both kicks is .45. If Bill is successful on the first kick, what is the possibility that he makes the second kick? Because the success of the second kick is dependent on the outcome of the first kick, we get:

$$P(\text{2nd kick/1st kick}) = \frac{P(\text{1st kick and 2nd kick})}{P(\text{1st kick})}$$

$$= \frac{.45}{.70} = .6429 = 64.29\%$$

APPLIED COUNTING TECHNIQUES

To calculate the probability of any simple event, it is necessary to obtain both the total number of outcomes and the number of that total that are favorable to the outcome of an experiment. To arrive at these numbers when they are large and difficult to count, there are 3 essential counting techniques to learn.

Basic Counting Principle

When an experiment (i.e., rolling one die) can be performed in x ways and a corresponding experiment (i.e., rolling a second die) can be performed in y ways, the combined experiment (rolling two dice) can be performed in $x \cdot y$ (6 · 6) ways.

Examples

1) A menu at your favorite restaurant contains 6 appetizers, 3 soups, 10 entrees, and 5 desserts. On Thursdays, for $7.95 you have your choice of an appetizer, a bowl of soup, an entree, and a dessert. How many total dinners can be arranged on Thursday nights so that no two are duplicated exactly?

$$\underset{\text{appetizers}}{6} \times \underset{\text{soups}}{3} \times \underset{\text{entrees}}{10} \times \underset{\text{desserts}}{5} = 900 \text{ total dinners}$$

2) If Peter, Mary, and Paul take a bus that makes 12 stops from where they work to the beach and they each get off at a different spot, how many total ways are there for them to leave the bus?

$$\underset{\text{12 ways}}{\text{Peter}} \times \underset{\text{11 ways}}{\text{Mary}} \times \underset{\text{10 ways}}{\text{Paul}} = 1320 \text{ ways to leave the bus}$$

Permutations

Permutations are *ordered* arrangements. Order is the key to determining when to use permutations to arrive at the total number of outcomes. When switching the order of the objects used in the experiment produces a unique arrangement, we are going to use the formula for permutations:

$$n Pr = n(n-1) \cdots (n-r+1) = \frac{n!}{(n-r)!}$$

where,

- n = total number of objects
- r = number of ways to arrange n objects
- ! = factoral (i.e., 4! = 4 · 3 · 2 · 1 and 3! = 3 · 2 · 1)

NOTE: The left side of the formula is really only a symbol, but it is this symbol that is important. If you make circles to represent r, the number of ways to arrange n objects, then all you have to do is fill in the number of n objects that fits in each circle and multiply the numbers to get the answer. Using the letters "STERN", we can rearrange them to form 2-, 3-, 4-, or 5-letter arrangements. Let's say we want to arrange the 5 letters into 4-letter groupings. Using the symbol $n Pr,$ $n = 5$ and $r = 4$. We start by drawing 4 circles.

⑤

Since we can use any of the letters for the first circle, we put the number 5 in the first circle.

⑤　　　④

We now have 4 letters available for the second circle.

⑤　　④　　③

For the third circle, we have 3 letters left.

⑤　　④　　③　　②

That leaves us 2 letters for the fourth circle. We now multiply 5 × 4 × 3 × 2 and we have 120 ways to arrange 5 objects in 4 ways.

This will work for any 5 objects, whether they might be people, cars, colors, countries, letters, etc.

Using the formula, we would have gotten:

$$n Pr = \frac{n!}{(n-r)!} = \frac{5!}{(5-4)!} = \frac{5 \cdot 4 \cdot 3 \cdot 2 \cdot 1}{1} = 120$$

The formula works okay for the total number of objects, but when we solve for the number of favorable objects, my system (tried and true) makes it easier to visualize the experiment. Using the same 5 letters, "STERN", what is the probability the ordered arrangement will begin with the letter "R"?

Because of the limitation on the first circle which can only be "R", we put a 1 in the first circle:

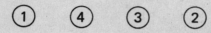

We then have 4 letters for the second circle, then 3 letters for the third and two for the fourth. 1 X 4 X 3 X 2 = 24 favorable outcomes, or the probability that the arrangement will begin with an "R" is $\frac{24}{120}$.

Examples

1) There are seven students (Mary, Debbie, Liz, Susan, Bob, Frank, and John) participating in elections for 4 student body offices: President, Vice-President, Secretary, and Treasurer.

   i) What is the probability that a girl becomes President and a boy Vice-President?

   ii) What is the probability that Bob becomes President and girls win the other 3 offices?

     a) The fist step is to determine the total number of possible arrangements for 7 students in 4 offices.

$$nPr = \frac{7!}{(7-4)!} = \frac{7!}{3!} = \frac{7 \cdot 6 \cdot 5 \cdot 4 \cdot \cancel{3} \cdot \cancel{2} \cdot \cancel{1}}{\cancel{3} \cdot \cancel{2} \cdot \cancel{1}} = 840$$

     b) To find the number of favorable outcomes for (a) we draw our 4 circles.

In the question you are told the 1st spot (President) must be a girl. There are 4 girls and any one can be President. Put a 4 in the 1st spot.

c) The 2nd spot (Vice-President) must be a boy. There are 3 boys, any boy can be Vice-President. Put a 3 in the 2nd spot.

P	VP	T
④	③	⑤

d) There are no limitations on the last 2 spots and there are 5 students left. Any of the 5 can fill that spot. Put a 5 in the 3rd spot.

P	VP	T	S
④	③	⑤	④

e) That leaves us with 4 people to fill the last spot. We then multiply 4 × 3 × 5 × 4 and get 240 favorable outcomes. The probability $= \frac{240}{840}$.

f) To find the number of favorable outcomes for ii), we again use our 4 circles.

P	VP	T	S
①	④	③	②

g) The 1 in the first spot represents Bob. The 4, 3, and 2 in the other spots represent the manner 4 girls are arranged 3 ways. By multiplying 1 × 4 × 3 × 2, we get 24 favorable outcomes. The probability $= \frac{24}{840}$.

2) The letters of the name "DEBRA" are arranged in all possible ways. If we choose at random, what is the probability our 5-letter arrangement will:

i) Begin with a vowel and end with a consonant?

a) The total number of arrangements (the denominator of our probability ratio) is:

$$n Pr = 5P5 = \frac{5!}{(5-5)!} = \frac{5!}{0!} = 5 \cdot 4 \cdot 3 \cdot 2 \cdot 1 = 120$$

b) NOTE: 0! = 1. The numerator of the ratio, the number of favorable outcomes is:

②	③	②	①	③

c) By multiplying 2 × 3 × 2 × 1 × 3 and placing the result above 120, we get: $\frac{36}{120}$

ii) Be "BREAD"?

a) The total number of arrangements is the same—120 arrangements.

b) The number of favorable outcomes = 1 because there is a limitation on all 5 spaces.

Ⓑ	Ⓡ	Ⓔ	Ⓐ	Ⓓ

Probability $= \frac{1}{120}$

Combinations

When the order of an arrangement is not significant, such as arranging 5 identical orange balls or generic groups of people (i.e., Democrats and Republicans), we refer to the arrangement as a combination, not a permutation. The formula for combination is:

$$\frac{n}{r} = \frac{n(n-1)\cdots(n-r+1)}{r!} = \frac{n!}{r!(n-r)!}$$

where,

n = total number of objects
r = number of ways to arrange n objects
! = factoral (0! = 1)

Examples

1) $\dfrac{9}{6} = \dfrac{9!}{6!(9-6)!} = \dfrac{9 \cdot 8 \cdot 7 \cdot 6 \cdot 5 \cdot 4 \cdot 3 \cdot 2 \cdot 1}{6 \cdot 5 \cdot 4 \cdot 3 \cdot 2 \cdot 1 \cdot 3 \cdot 2 \cdot 1} = \dfrac{9 \cdot 8 \cdot 7}{3 \cdot 2 \cdot 1} = 84$

2) $\dfrac{7}{5} = \dfrac{7!}{5!(7-5)!} = \dfrac{7 \cdot 6 \cdot 5 \cdot 4 \cdot 3 \cdot 2 \cdot 1}{5 \cdot 4 \cdot 3 \cdot 2 \cdot 1 \cdot 2 \cdot 1} = \dfrac{7 \cdot 6}{2 \cdot 1} = 21$

Once again, when solving a probability problem involving combinations, you must calculate the total number of outcomes first. Use the preceding formula and solve for n using all the objects presented. The number of favorable outcomes, depending on a particular experiment, will vary according to the limitations placed on the data.

Examples

1) From a group of 6 men and 5 women, how many 5-person basketball teams can be formed?

   i) Because the order of selection and the positions on the team do not matter, we solve this problem using combinations:

   ii) $\dfrac{n}{r} = \dfrac{11}{5}$ (men + women) $= \dfrac{11!}{5!(11-5)!}$

   $$= \frac{11 \cdot 10 \cdot 9 \cdot 8 \cdot 7 \cdot 6 \cdot 5 \cdot 4 \cdot 3 \cdot 2 \cdot 1}{5 \cdot 4 \cdot 3 \cdot 2 \cdot 1 \cdot 6 \cdot 5 \cdot 4 \cdot 3 \cdot 2 \cdot 1}$$

   $$= \frac{11 \cdot 10 \cdot 9 \cdot 8 \cdot 7}{5 \cdot 4 \cdot 3 \cdot 2 \cdot 1}$$

   $$= 462 \text{ combinations}$$

2) What is the probability that based on the information in 1), a team will have 3 men and 2 women? In this case, the number of favorable outcomes is divided between men and women. We still use all 11

people and the team will still consist of 5 men and women. Using the symbol for combinations because the order of selection is not important, we get:

$$\frac{\binom{6}{3}\binom{5}{2}}{462} = \frac{\dfrac{6!}{3!(6-3)!}\dfrac{5!}{2!(5-2)!}}{462}$$

$$= \frac{\left(\dfrac{6\cdot5\cdot4\cdot3\cdot2\cdot1}{3\cdot2\cdot1\cdot3\cdot2\cdot1}\right)\left(\dfrac{5\cdot4\cdot3\cdot2\cdot1}{2\cdot1\cdot3\cdot2\cdot1}\right)}{462}$$

$$= \frac{\left(\dfrac{6\cdot5\cdot4}{3\cdot2\cdot1}\right)\left(\dfrac{5\cdot4}{2\cdot1}\right)}{462} = \frac{(20)(10)}{462} = \frac{200}{462}$$

3) Using the same 11 people, what is the probability of at least 4 men on the team?

At least 4 men means that there could be 4 men and 1 woman or 5 men. When there are 2 probabilities, we solve for each and add the probabilities. Thus:

$$P(4\text{ or }5\text{ men}) = \frac{\binom{6}{4}\binom{5}{1} + \binom{6}{5}\binom{5}{0}}{\binom{11}{5}}$$

$$= \frac{\left(\dfrac{6\cdot5\cdot4\cdot3\cdot2\cdot1}{4\cdot3\cdot2\cdot1\cdot2\cdot1}\right)\left(\dfrac{5}{1}\right) + \left(\dfrac{6\cdot5\cdot4\cdot3\cdot2\cdot1}{5\cdot4\cdot3\cdot2\cdot1\cdot1}\right)(1)}{462}$$

$$= \frac{\left(\dfrac{6\cdot5\cdot4}{2\cdot1}\right)(5) + (6)(1)}{462}$$

$$= \frac{300+6}{462} = \frac{306}{462}$$

THE BINOMIAL DISTRIBUTION

A discrete distribution, the binomial distribution of probability, involves a fixed number of experiments (trials). Each of these experiments (i.e., flipping a coin) is independent and there is a situation that either occurs or does not occur on each trial. In this section we are interested in the probability distribution of the number of successes. Using the lower case p to denote success in an experiment, you will learn to calculate the possible number of successes and their corresponding probabilities.

Using a practical and versatile statistical tool, the tree diagram (aptly named, because without it you might just be up one), we employ a problem that considers the probability of selecting, at random, a lemon or an orange from a box containing 4 lemons and 2 oranges. Figure 4-C illustrates all the possible outcomes and their probabilities when 3 random selections are made. What is the probability of 3 lemons? Of 2 lemons and 1 orange?

If you've been paying close attention, you will remember that the probability of selecting a lemon on one trial is $\frac{4}{6}$ or $\frac{2}{3}$. The probability of selecting an orange is $\frac{2}{6}$ or $\frac{1}{3}$. The total probability, $\frac{2}{3} + \frac{1}{3}$, is 1. If the selected fruit is replaced, the probabilities will remain the same on all following random selections.

In our tree diagram, "L" will represent the selection of a lemon and "O" an orange.

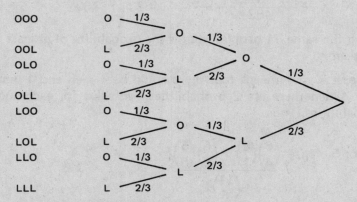

Figure 4-C. Tree diagram for 3 trials.

In Figure 4-C, the tree branches represent all the possible outcomes of the experiment which involves 3 separate random trials. By counting the number of branches at the left-hand side of our diagram, you can see that there are 8 outcomes.

NOTE: Total number of outcomes = $(n)^r$
where,

n = number of outcomes in an experiment
r = number of trials

From our diagrammed problem we get,

$$(n)^r = (2)^3 = 2 \times 2 \times 2 = 8 \text{ outcomes}$$

From our discussion of independent events in this chapter (page 43), the probabilities of any 3 outcomes can be computed by multiplication. For example:

$$P(\text{lemon, orange, lemon}) = \frac{2}{3} \times \frac{1}{3} \times \frac{2}{3} = \frac{4}{27}$$

For any 3 outcomes in our experiment, the total number of outcomes will always be 27. The number of favorable outcomes will vary according to the particular experiment.

$$P(\text{orange, orange, orange}) = \frac{1}{3} \times \frac{1}{3} \times \frac{1}{3} = \frac{1}{27}$$

The 8 outcomes (based on 3 trials) in our lemon and orange test are mutually exclusive and exhaustive. They are mutually exclusive because there is only 1 possible outcome for each selection (lemon-oranges don't exist); they are exhaustive because the 8 outcomes represent the entire sample space. As a result, the probability of any event involving a combination of outcomes is calculated by adding the appropriate probabilities that correspond to the particular experiment.

EXAMPLES

1) What is the probability of obtaining 3 lemons in 3 random selections?

Because the only outcome that corresponds to the experiment is (L, L, L), the probability $= \frac{2}{3} \times \frac{2}{3} \times \frac{2}{3} = \frac{8}{27}$.

2) What is the probability of obtaining 2 lemons and 1 orange?
   i) 3 outcomes correspond to obtaining 2 lemons. They are (L, L, O), (L, O, L) and (O, L, L). We then add the probabilities and get:
   ii) P(2 lemons) = P(L, L, O) + P(L, O, L) + P(O, L, L)

$$= \frac{2}{3} \times \frac{2}{3} \times \frac{1}{3} + \frac{2}{3} \times \frac{1}{3} \times \frac{2}{3} + \frac{1}{3} \times \frac{2}{3} \times \frac{2}{3}$$

$$= \frac{4}{27} + \frac{4}{27} + \frac{4}{27} = \frac{12}{27}$$

3) What is the probability of obtaining 1 lemon?
   i) 3 outcomes correspond to obtaining 1 lemon. They are (L, O, O), (O, L, O), and (O, O, L). We then add the probabilities and get:
   ii) P(1 orange) = P(L, O, O) + P(O, L, O) + P(O, O, L)

$$= \frac{2}{3} \times \frac{1}{3} \times \frac{1}{3} + \frac{1}{3} \times \frac{2}{3} \times \frac{1}{3} + \frac{1}{3} \times \frac{1}{3} \times \frac{2}{3}$$

$$= \frac{2}{27} + \frac{2}{27} + \frac{2}{27} = \frac{6}{27}$$

What we've just covered is an example of a *binomial probability experiment* in which the probability of a successful outcome (p) is the same in every trial. The probability of failure is therefore $1 - p$.

In general, the number of successes in a binomial experiment are impossible to predict because the outcomes are the result of chance in random selections. The number of successes (we will use "x") is the random variable in a binomial experiment. Instead of trying to prepare a tree diagram which, although fun, is time-consuming, we will use a general formula.

$$\text{Probability } (x \text{ successes}) = \frac{n!}{x!(n-x)!} \cdot p^x (1-p)^{n-x}$$

where,
   n = number of independent trials
   p = probability of success on each trial
   x = any integer 0, 1, 2, 3, 4, $\cdots n$

Employing this formula in our experiment with lemons and oranges, we get:

$$\begin{aligned}
\text{Probability (3 lemons)} &= \frac{3!}{3!(3-3)!} \times \left(\frac{2}{3}\right)^3 \left(1 - \frac{2}{3}\right)^{3-3} \\
&= \frac{3 \cdot 2 \cdot 1}{3 \cdot 2 \cdot 1 \,(1)} \times \left(\frac{2}{3}\right)^3 \left(\frac{1}{3}\right)^0 \\
&= 1 \times \left(\frac{2}{3}\right)^3 \times (1) \\
&= 1 \times \frac{8}{27} \times 1 \\
&= \frac{8}{27}
\end{aligned}$$

This probability is denoted as: $b(x; n, p)$.
The mean $(\bar{x}) = (n \cdot p)$
The standard deviation $(\sigma) = \sqrt{np\,(1-p)}$
   To solve the following examples, we must refer to Table 4-A on page 44. (Using the formula, the work can also be done by hand.)

EXAMPLES

1)   What is the probability of success when:

$$n = 8, p = .02, \text{ and } x = 4?$$
$$b = (4; 8, .02) = 0.046$$

2)   What is the probability for success when:

$$n = 10, p = .6, \text{ and } x = 5?$$
$$b = (5; 10, .6) = .201.$$

3) When a test was given, the probability of getting an "A" was .70. If 10 students took the test:

    i) What is the expected number of students who will get an "A"?

        The expected number is the mean.

$$\text{Mean } (\bar{x}) = n \cdot p = 10\,(.70) = 7 \text{ students}$$

    ii) What is the standard deviation of the number of students taking the test?

$$
\begin{aligned}
\text{Standard deviation } (\sigma) &= \sqrt{np(1-p)} \\
&= \sqrt{10(.7)(1-.7)} \\
&= \sqrt{7\,(.3)} \\
&= \sqrt{2.1} = 1.45
\end{aligned}
$$

Well, that's it for your lessons on probability theory. It's a lot to grasp, but after the following exercises, the use of various formulas in various situations will make more and more sense. Just remember that it's not necessary to memorize the formulas. Use them; use the examples to do the problems one step at a time. The probability for success is good.

EXERCISES

Exercise 4-1: Consider a deck of 52 cards. What is the probability of the following?

a) selecting a king
b) selecting either a king or a queen
c) selecting a king or a club
d) selecting an ace, king, queen, jack, and ten in succession without replacement of cards

Exercise 4-2: Consider 2 independent events, A and B. P(A) = .9 and P(*not* B) = .2. Solve for:

a) P(A or B)
b) P(A and B)
c) P(not A or B)
d) P(not A and not B)

Exercise 4-3: The student body government of S. Stern High School in Brooklyn is composed of 4 sophomores, 5 juniors, and 6 seniors. What is the probability that a committee of 6 students will contain the following?

a) 2 sophomores
b) 2 juniors and 3 seniors
c) at least 4 seniors

Exercise 4-4: Peggy, Joan, Mary, Susan, Frank, Ken, Steve, and Bill are competing for 1st, 2nd, and 3rd prizes in an art competition. What is the probability that:

a) Joan or Mary wins 2nd prize?
b) the girls win all the prizes?
c) a boy wins 1st and 3rd prizes?

Exercise 4-5: Consider the rolling of a single die 6 consecutive times. Use the bionomial probability formula to determine the probability of rolling the number 5:

a) no times
b) 3 times

Exercise 4-6: Lovers of Ray's Pizza in New York brag that they can identify their favorite pizza by tasting it. From experience the probability of selecting the correct pizza is .45. If 10 people take the taste test, calculate (using the binomial formula):

a) the probability that 5 people will pick Ray's Pizza
b) the expected number of people who will guess correctly
c) the standard deviation of the people guessing correctly

Take several deep breaths, holding the last for 10 seconds. After exhaling slowly, turn the page and be glad for Chapter 5. The worst is over . . . I think.

5

PROBABILITY DISTRIBUTIONS

"The Behavior of Random Variables" or "Should I Buy That Earthquake Insurance?"

As the counterpart of the frequency distribution, the probability distribution classifies discrete and continuous data according to the relative frequency (probability) of an event or experiment. In this chapter we will be discussing and using random variables in our problem solving.

Random variables are assigned numerical values given to the outcomes of a probability (chance) experiment. This enables us to obtain theoretical values for the mean, the variance, and the standard deviation of a particular distribution using the probability of a variable $[p(x)]$ rather than the frequency (f) in our formulas.

DISCRETE RANDOM VARIABLES

Consider a probability experiment involving the number of states voting on a constitutional amendment. There might be 0, 1, 2, 3, and so on, up to 52 states voting for passage. Because there are breaks between the values (only whole numbers, no infinite fractions) the random variable is said to be discrete.

In order to show the probability function, instead of the frequency function of a random variable, each random variable in a distribution is assigned a corresponding probability. The total of these probability functions is *always* equal to 1.

$$\sum_{i=1}^{n} p(x) = 1$$

Consider that a random variable (number of hurricanes in Florida) has the following probability function:

x	0	1	2	3	4	5
p(x)	.05	.15	.25	.25	.20	.10

If you were an orange grower and wanted to know the probability of no more than 3 hurricanes, you would then calculate the probability $(x \leq 3)$. $x \leq 3$ means that x is less than or equal to 3. We get:

$$P(x \leq 3) = p(x = 0) + p(x = 1) + p(x = 2) + p(x = 3)$$
$$= .05 + .15 + .25 + .25$$
$$= .70$$

Depending on the question, we add the appropriate probabilities.

EXAMPLES

Using the following probability function for the number of customers per hour in a neighborhood bank, what is the probability that:

1) $x \geq 2$
2) $4 > x \geq 1$ or $1 \leq x < 4$

x	0	1	2	3	4	5
$p(x)$	.03	.09	.21	.25	.27	.15

i) $p(x \geq 2) = p(x = 2) + p(x = 3) + p(x = 4) + p(x = 5)$
     $= .21 + .25 + .27 + .15$
     $= .88$
ii) $4 > p(x) \geq 1 = p(x = 3) + p(x = 2) + p(x = 1)$
     $= .25 + .21 + .09$
     $= .55$

To solve probability functions for the mean, the variance, and the standard deviation, it is first necessary to place our horizontal function into vertical form to complete the distribution. Once again, we will use the probability function for the number of hurricanes in Florida.

Table 5-A. Probability Distribution of Hurricanes in Florida

x	$p(x)$	$p(x) \cdot x$	$p(x) \cdot x^2$
0	.05	0	0
1	.15	.15	.15
2	.25	.50	1.00
3	.25	.75	2.25
4	.20	.80	3.20
5	.10	.50	2.50
Total	1.00	2.70	9.10

NOTE: Our formulas for calculating the mean, the variance, and the standard deviation remain the same (as in Chapter 3) *except* that we substitute $p(x)$ for f and n, the total frequency, is not necessary because the total probability always equals 1. Ergo, relative frequency, not the frequency, becomes the standard in the following:

a) Mean $(\bar{x}) = \sum_{i=1}^{n} p(x) \cdot x$

b) Variance $(\sigma^2) = \sum_{i=1}^{n} p(x) \cdot x^2 - (\mu)^2$, where μ = probability mean

c) Standard deviation $(\sigma) = \sqrt{\sum_{i=1}^{n} p(x) \cdot x^2 - (\mu)^2}$

Solving for Figure 5-A, we get:

a) Mean $(\bar{x}) = \sum_{i=1}^{n} p(x) \cdot x = 2.70$ expected hurricanes

b) Variance $(\sigma^2) = \sum_{i=1}^{n} p(x) \cdot x^2 - (\mu)^2$

$$= 9.10 - (2.70)^2$$
$$= 9.10 - 7.29$$
$$= 1.81 \text{ hurricanes squared}$$

c) Standard deviation $(\sigma) = \sqrt{1.81}$ hurricanes squared
$$= 1.345 \text{ hurricanes}$$

Let us again consider the situation pictured in Figure 4-C involving the selection of lemons or oranges. For the sake of our probability function, the random variable will be the number of lemons selected in 3 trials.

x	0	1	2	3
$p(x)$	$\dfrac{1}{27}$	$\dfrac{6}{27}$	$\dfrac{12}{27}$	$\dfrac{8}{27}$

x	$p(x)$	$p(x) \cdot x$	$p(x) \cdot x^2$
0	$\dfrac{1}{27}$	0	0
1	$\dfrac{6}{27}$	$\dfrac{6}{27}$	$\dfrac{6}{27}$
2	$\dfrac{12}{27}$	$\dfrac{24}{27}$	$\dfrac{48}{27}$
3	$\dfrac{8}{27}$	$\dfrac{24}{27}$	$\dfrac{72}{27}$
	$\dfrac{27}{27}$	$\dfrac{54}{27} = 2$	$\dfrac{126}{27} = 4\dfrac{2}{3}$

i) $\mu = p(x) \cdot x = \dfrac{54}{27} = 2$

ii) $\sigma^2 = p(x) \cdot x^2 - (\mu)^2$

$$= 4\dfrac{2}{3} - (2)^2$$

$$= 4\dfrac{2}{3} - 4$$

$$= \dfrac{2}{3} = .667$$

iii) $\sigma = \sqrt{.667} = .817$

NOTE: When the number within the square root sign ($\sqrt{}$) is less than 1, the square root will be a greater quantity. When the number within the square root sign is greater than 1, its square root will be a smaller number, such as,

$$\sqrt{.5} = .707$$
$$\sqrt{1.5} = 1.225$$

CONTINUOUS RANDOM VARIABLES

When dealing with variables that can assume any values in the interval (no breaks), such as length measured in miles, yards, feet, inches, and fractions of inches, it is necessary to graph the data rather than use a probability distribution table. Because the possible random variables are infinite (no sheet of paper has a chance to hold them all), we use a probability density curve to pictorially display the range of data.

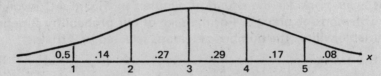

Figure 5-A. Probability density curve for the life span (in years) for pet turtles.

NOTE: For a probability density curve to be accurate in its description of the probability function of the random continuous variable, it must

1) always be above the X-axis (horizontal axis), because there cannot be a negative probability.
2) give the probabilities of a specific range (area) of variables.
3) show that the area (total probability) above the X-axis is always equal to 1.
4) not supply any of the continuous random variables with a specific value.

EXAMPLES

1) Using the probability density curve in Figure 5-A, solve for the following:

   i) $p(x > 1)$. The probability that $x > 1$ is equal to the total probability to the right of 1. Probability $(x > 1) = .14 + .27 + .29 + .17 + .08 = .95$

   ii) $P(1 < x < 4)$. The probability that x is between 1 and 4 $= .14 + .27 + .29 = .73$

   iii) $P(not\ 1 < x < 4)$
        $P(not\ x < x < 4) = 1 -$ probability $(1 < x < 4)$
        $= 1 - .73 = .27$

2)

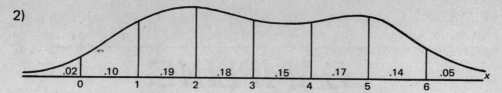

Figure 5-B. Probability density curve.

i) What is the probability of the following in Figure 5-B:
$x < 4$ P($x < 4$) equals everything to the left of 4
P($x < 4$) = .15 + .18 + .19 + .10 + .02 = .64

ii) What is the probability of the following:
$1 < x < 4$
Probability between 4 and 1 = .19 + .18 + .15 = .52

EXERCISES

Exercise 5-1: The amount of money a certain freelance writer earns is shown in the following probability function:

(amount earned) x	1000	1200	1600	2000	2400
$p(x)$	.20	.22	.24	.21	.13

a) What is the probability that this writer will earn more than $1300? . . . less than $2100?
b) Find the mean, variance, and standard deviation of the amount of money earned.

Exercise 5-2:

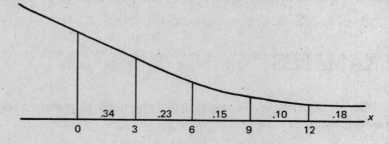

.34   .23   .15   .10   .18

0   3   6   9   12

Using the preceding probability density for the useful life (in months) of light bulbs, answer the following questions:

a) What is the probability that a light bulb will last more than 6 months?
b) What is the probability that a light bulb will last between 3 and 12 months?

At this point, we should stop to ask ourselves how well we understand the material presented thus far. The following chapters are longer and more complex, and they build upon the material in Chapters 1 to 5.

If you are ready, we will proceed.

6

SAMPLING DISTRIBUTIONS

"The Normal Curve" or "Is That Your Most Educated Guess?"

In this chapter, you will become acquainted with probability distributions of certain statistics that are computed from sample data; hence the name, sampling distribution.

From parameters derived from a population—the expectation (mean) and the dispersion (variance)—we will address our attention to comparing the distribution (spread) of the mean in selected samples.

Underlying our study of the sampling distribution of the mean and other sampling distributions is the need to understand the normal distribution and the normal curve.

THE NORMAL DISTRIBUTION

The normal distribution, based on a *continuous* random variable, is always described by a probability density curve. The density curve is used to measure the variable which can (theoretically) take on any value.

EXAMPLES

Continuous variables result from the need to measure variables such as the following:

1) The weights of cars, people, food, etc.
2) The distances traveled by cars in braking.
3) The lengths of people's feet, hair, and patience.
4) The volume of liquid in a glass, tank, or reservoir.
5) The time it takes for water to boil, for your wife (or husband) to get ready, or for a train to arrive.

The probability density curve to describe a normal distribution is symmetrical on both sides of the mean. In a special case, when we have a mean of 0 and a standard deviation of 1, we have a standard normal curve.

A random variable with a mean $(\mu) = 0$ and a standard deviation $(\sigma) = 1$ is said to be a *standard normal variable*.

$$\left[\frac{\overline{x} - \mu}{\sigma}\right]$$ represents the standard normal variable

NOTE: The letter z is used to denote this random variable. Appendix Table 2 (page 126) will enable us to find the probabilities found under the standard normal curve.

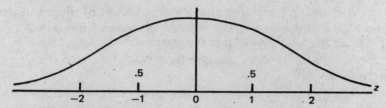

Figure 6-A. Standard normal curve.

NOTE: The three keys to solving problems involving normal distributions described by normal curves are the following:

1) Because the curve is symmetrical, the probability of an event is 0.5 on either side of the mean for a total probability of 1.0.
2) Negative differences between the population mean and the sample mean do not matter. We are concerned only with absolute values which indicate the dispersion (+ and −) from the mean (μ).
3) Approximately 99.7% of the area of a normal distribution is contained within 3 standard deviations of the mean $(-3 \leq z \leq +3)$.

EXAMPLES

Using Appendix Table 2 (page 126), find the following probabilities when z is a standard normal variable:

1) $p\,(z \leq 1.75)$.
   i) The probability that the standard normal variable is less than or equal to 1.75 includes the total area under the curve to the left of 1.75.

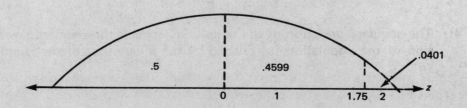

ii) This includes the probability to the right of the (0) and 1.75 plus the probability to the left of the mean:

Using Appendix Table 2:

$$p(z \leq 1.75) = .5 + .4599 = .9599 = 95.99\%$$
$$p(z \geq 1.75) = .5 - .4599 = .0401 = 4.01\%$$
$$.9599 + .0401 = 1.000 = 100\%$$

2) $P(-1.8 < z < 2.1)$.
  i) The probability that the standard normal variable is between -1.8 and 2.1 includes the area to the left of the mean from 0 to -1.8 and the area to the right of the mean from 0 to 2.1.
  ii) $P(-1.8 < z < 2.1) = .4641 + .4821$
  $$= .9462 = 94.62\%$$

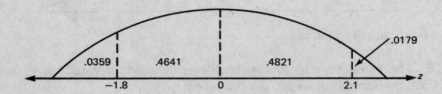

SAMPLING DISTRIBUTION OF THE MEAN

Employing our knowledge of the normal curve, we are now ready to solve problems involving theoretical sampling distributions. In these cases, the mean of a population and its standard deviation are used to formulate probabilities for individual samples. The *Central Limit Theorem* provides the mathematical basis for describing the nature of a sampling distribution. As the size of the sample (n) increases, the sampling distribution of the mean ($\bar{x}$) has the following properties:

1) The mean of the distribution (μ) is located at the center of the normal curve.
2) The mean of the sample means approximates the mean of the population. This produces the symmetry of the distribution.
3) The variance of the sample means ($\sigma_{\bar{x}}^2$) divided by the size of the sample (n).

$$\sigma_{\bar{x}}^2 = \frac{\sigma^2}{n}$$

4) The standard deviation of the sample ($\sigma_{\bar{x}}$) equals the standard deviation of the population (σ) divided by the square root of the sample (n).

$$\sigma_{\bar{x}} = \frac{\sigma}{\sqrt{n}}$$

The standard deviation of the sampling distribution is also known as *the standard margin of error*.

NOTE: To calculate the areas between z scores in a normal curve, add the areas if the z scores are on opposite sides of the mean. If the z scores are on the same side of the mean, subtract the area of the z score that is nearer the mean from the z score farther from the mean.

The arithmetic mean ($\bar{x}$) as explained in Chapter 3 is a descriptive statistic of a sample of data. However, the mean of a population (μ) is a parameter. Conclusions about the parameter of a population (a descriptive measure) are based on a sample statistic. Subject to limitations, a researcher is able to make certain inferences about population parameters from an understanding of the sample statistics. This connective reasoning is called *inferential statistics*. In this chapter, we will consider the relationship between sampling and probability theory in addition to the concept of the underlying normal distribution of all conceivable outcomes.

EXAMPLES

1) Given a population with a mean (μ) = 150 and a standard deviation (σ) of 17.5, what is the probability that a sample of 100 will have a mean greater than 152?

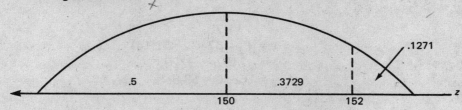

i) The first step is to correlate the standard deviation of the sample.

$$\sigma_{\bar{x}} = \frac{\sigma}{\sqrt{n}} = \frac{17.5}{\sqrt{100}} = \frac{17.5}{10} = 1.75$$

ii) We then compare the sample mean with the mean of the population and divide by the standard deviation of the sample ($\sigma_{\bar{x}}$).

$$P(z > 152) = P\left(z > \frac{152 - 150}{1.75}\right)$$
$$= P\left(z > \frac{2}{1.75}\right)$$
$$= P(z > 1.14)$$

iii) Using Appendix Table 2 (page 126), we get

$$P(z > 152) = .5 - .3729$$
$$P(z > 152) = .1271 = 12.71\%$$

2) Using the data in Example 1), what is the probability that a sample will have a mean greater than 149?

   i) $P(z > 149) = P\left(z > \dfrac{149 - 150}{1.75}\right)$

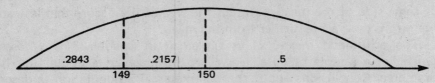

   ii) $P(z > 149) = P\left(z > \dfrac{-1}{1.75}\right)$

   $= P(z > -.57)$

   $= .2157 + .5$

   $= .7157 = 71.57\%$

3) What is the probability that the sample mean will be between 149 and 152?

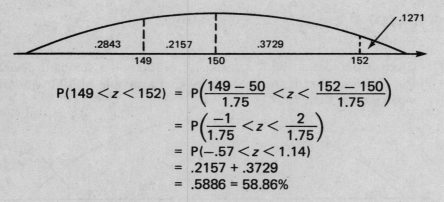

   $P(149 < z < 152) = P\left(\dfrac{149 - 50}{1.75} < z < \dfrac{152 - 150}{1.75}\right)$

   $= P\left(\dfrac{-1}{1.75} < z < \dfrac{2}{1.75}\right)$

   $= P(-.57 < z < 1.14)$

   $= .2157 + .3729$

   $= .5886 = 58.86\%$

OTHER SAMPLING DISTRIBUTIONS

In this section, we will learn about 3 additional distributions for continuous random variables and the use of 3 additional and important tables.

The Student's τ-Distribution

In testing, when the variance of the population is unknown and it is necessary to use the variance of the sample to estimate it, the normal curve becomes inadequate for describing the sampling distribution of the mean. When a sample is small, the traditional statistical procedures that use the normal distribution as the sampling distribution come up short.

The Student's (a pen name for an Irish chemist, William S. Gossett) τ-distributions are an interrelated group of symmetrical distributions. As the sample size increases, the τ-distribution will increasingly approximate the normal distribution. Thus, there is a specific τ-distribution for every sample size. A τ-distribution is a continuous distribution that is symmetrical on both sides of the vertical axis (Y-axis) at 0. The appropriate τ-distribution is dependent on a parameter called the *degrees of freedom (df)*.

In computing the statistic τ, we assume the population variance and standard deviation are unknown and we use the sample variance (or standard deviation) as an estimate.

$$\tau = \frac{\bar{x} - \mu}{S_{\bar{x}}} \text{ and } S_{\bar{x}} = \frac{S}{\sqrt{n}}$$

where

τ = test statistic
$\bar{x}$ = sample mean
μ = population mean
$S_{\bar{x}}$ = standard deviation of sample
n = sample size

NOTE: Fundamentally mathematical in nature, the degrees of freedom is dependent upon the number of sample observations (n). It is the number of observations minus the number of restrictions placed upon them. The distributions in Appendix Table 3 (page 127) are in standard-score form with a mean equal to 0 and a standard deviation of 1.

NOTE: The τ-distribution for infinite (∞) degrees of freedom is identical to the normal distribution. Each row of Appendix Table 3 (page 127) represents a unique τ-distribution and each distribution is associated with a unique number of degrees of freedom. The column headings in the table—.0005, .005, .01, .025, .05, .10—represent the proportion of the area remaining in the tails of the described distribution (*see* Figure 6-B).

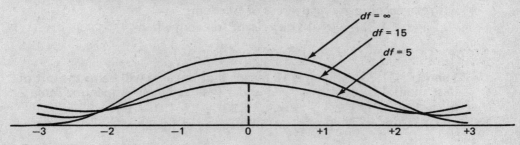

Figure 6-B.

The proportions in the tails of the distribution represent the areas less than −3 standard deviations and greater than +3 standard deviations about the mean.

Examples

1) Using Appendix Table 3 (page 127), define:
   i) $\tau_{\gamma,a}$ for τ 25, .025.

a) $\tau_{\gamma,\alpha}$ is defined as a τ value with γ degrees of freedom such that an area α is in the right tail of the distribution. From Appendix Table 2 (page 126), we get:

$$\tau_{25,\,.025} = 2.060$$

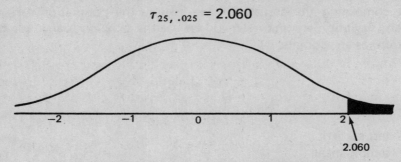

b) The shaded area to the right of 2.060 is .025.

ii) $\tau_{25,.975}$. Because a τ-distribution is symmetrical, $\tau_{25,.075}$ represents the negative of the τ-value $\tau_{25,.025}$. Therefore, $\tau_{25,.975} = -\tau_{25,.025}$ meaning the area covered will be in the left tail.

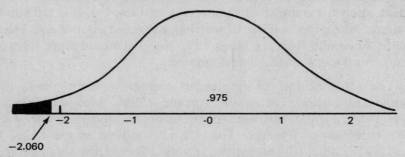

2) Find the probability when the test statistic (τ) has the τ-distribution with the accompanying degrees of freedom:

i) $P(\tau < 1.33)$, with 18 degrees of freedom when

$$\tau = \frac{x - \mu}{S_x}$$

The $P(\tau < 1.33)$ when $\gamma = 18$ means that the area will be to the left of (less than) 1.33 which is given under the heading in Appendix Table 2 (page 126) of .10. We get $P(\tau < 1.330) = 1 - .10 - .90$ or 90%.

ii) $P(-1.895 < \tau < 2.365)$ with 7 degrees of freedom.

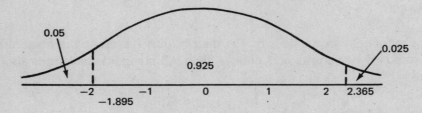

a) The probability is calculated by subtracting the area to the left of 1.895 and the right of 2.365 from 1.

b) Probability $(-1.895 < \tau < 2.365)$ with 7 *df* = $1 - 0.05 - 0.025 = 0.925$

The Chi-Square Distribution

The chi-square distribution is another distribution that describes a continuous random variable and depends upon the parameter, the number of degrees of freedom (df). However, this curve is always to the right of the vertical axis (Y-axis). When the number of degrees of freedom is small, the distribution is marked skewed (away from symmetry) to the right. As the df increases, the skewness disappears and the distribution appears more like a normal distribution. For a df greater than 30, the underlying distribution is the normal distribution. For a chi-square distribution, the statistics $X^2 = \frac{(n-1)s^2}{\sigma^2}$ has ($n-1$) degrees of freedom, with n the sample size.

NOTE: We will be using Appendix Table 4 (page 128) to solve chi-square distribution problems.

Examples

Use Appendix Table 4 (page 128) to define:

1) $x^2_{15,0.01}$

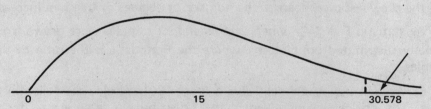

i) This curve is not symmetrical. Therefore, *all* of the obtained values come directly from Appendix Table 3 (page 127). $x^2_{15,0.01}$ is a value where the degrees of freedom = 15 and the area (α) = .01.

ii) From Appendix Table 4, we get:

$$x^2_{15,0.01} = 30.578$$

2) $x^2_{5,0.95}$.

i) The table for the chi-square (x^2) distribution does provide values for the area in the tail that is 0.095. From Appendix Table 3 (page 127), we get:

ii) $$x^2_{15,0.95} = 1.145$$

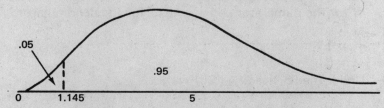

3) Calculate the probability if the test statistic has the chi-square distribution with 20 degrees of freedom.

    i)   Probability ($\mu < 9.237$).

    ii)   From Appendix Table 4 we find that the area to the right of 9.237 (with 20 df) is given by the column heading .98, the $P(\mu < 9.237)$ is equal to $1 - .98 = .02$

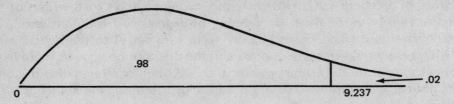

The f-Distribution

Like the Student's τ-distribution and the chi-square (x^2) distributions, the f-distributions are a family of distributions whose shape is dependent on a parameter. However the f-distributions, which are defined by *two* degrees of freedom values, are *not* symmetrical and have values which range from 0 to infinity (∞).

The f-distribution is skewed to the right but, like the chi-square distribution, the skewness disappears as the number of degrees of freedom increases.

The statistic $F = \dfrac{s_1^2}{s_2^2}$ when 2 independent samples were drawn from 2 normally distributed populations, where the F statistic is the ratio of these variables.

NOTE: We will use Appendix Table 5 (page 130) to calculate f-distribution values. The critical values given in the f-distribution table are all for the right-hand tail of the distribution. There are no values less than 1.0.

Examples

Use the f-distribution table (Appendix Table 5) to find:

1) $f_{15,8;.10}$
    i)   The f-distribution table coordinates 2 sets of degrees of freedom. By its definition, $f_{\gamma_1, \gamma_2; \alpha}$ denotes an f-distribution with γ_1 and γ_2 degrees of freedom leaving an area (α) in the right tail of the distribution.
    ii) In Appendix Table 4, $f_{15,8;.10}$ refers to the .10 heading where 15 df and 10 df intersect. That point = 2.46 (γ_1 represents the df for the numerator and γ_2 the df for the denominator.

2) $f_{11,6;.05}$
    i) $\gamma_1 = 11$
        $\gamma_2 = 6$
    ii) Under the 5% heading (.05) we get:

$$f_{11,6;.05} = 4.03$$

The three distributions introduced to you and explained in the preceding section, "Sampling Distribution of the Mean," will be employed in estimation and the testing of hypotheses in Chapters 7 and 8. When you next see them again, you will be fully capable of employing the Student's τ-distribution, the chi-square (x^2) distribution and the f-distribution to solve those problems presented. So, if any questions or confusion remains regarding the importance of these distributions, please be patient. I won't leave you out on a limb . . . too long.

EXERCISES

Exercise 6-1: Using the normal distribution, Appendix Table 2 (page 126):

a) The mean height of the men in the Danish Navy is 70 inches with a standard deviation of 2.5 inches. If 25 men recruits enter basic training, what is the probability (approximate) that the mean height of the men will be the following?
  i)   greater than 72 inches
  ii) less than 68.5 inches
  iii) between 68.5 and 72 inches

b) The expected time (mean) it takes a recruit to learn basic sailing skills is 39 days with a standard deviation of 2.1 days. With 25 new recruits, what is the probability that the mean learning time will be the following?
  i)   less than 37.5 days
  ii) between 38 and 40 days
  iii) greater than 41 days

Exercise 6-2: Using the Student's τ-distribution, Appendix Table 3 (page 127), find the probability if T (test statistic) has the τ-distribution with the given number df:

a) $P(T < 2.896)$ with 8 df
b) $P(-1.337 < T < 2.921)$ with 16 df

Exercise 6-3: Using the chi-square distribution, Appendix Table 4 (page 128), calculate the following probabilities if U (test statistics) has an x^2-distribution with the given number df:
a) $P(U < 2.733)$ with 8 df
b) $P(10.085 < U < 19.511)$ with 17 df

Exercise 6-4: Using the f-distribution, Appendix Table 5 (page 130), calculate:
a) $f_{20,9;0.25}$
b) $f_{7,4;0.01}$

7

ESTIMATION

"Determine Your Level of Competence" or "Your Guess Is Probably Better Than Mine"

Statistical Estimation is the process of approximating (estimating) a parameter from a corresponding sample statistic (characteristic). We will be working on developing 2 basic methods to obtain the rough ideas for our population parameters—point estimation and interval estimation.

NOTE: Because we can never know the population mean and, hence, the variance and standard deviation for certain unless we measure each member of a population, the process of estimation is part of our study of inferential statistics.

POINT ESTIMATION

We shall learn to provide "best" estimates for the population parameters mean (μ), variance (σ^2), and proportion (p). The guiding force in your arriving at the proper estimate is, believe it or not, your intuition.

EXAMPLES

If a population has a set of observations $x_1, x_2, \ldots$ to x_n, we then have:

1) The sample mean ($\bar{x}$) is a good point estimate for μ.

2) The sample variance ($\hat{s}^2$) = $\sum_{n-1}^{n} \frac{fx^2}{n-1} - (\bar{x})^2$ is a good point estimate of σ^2.

    NOTE: For the purpose of estimation we use $n - 1$ to represent the size of the sample.

3) When employing qualitative data—opinion surveys and evaluations—where we have x number of trials (experiments), then $\frac{x}{n}$ is a good point estimate of p.

Although the sample mean ($\bar{x}$) is apt to change from sample to sample in a given population, the mean ($\bar{x}$) as a point estimator refers to a mean computed from a given (random) set of sample values. If, from a population, you obtain the following set of values: 25, 29, 14, 18, 30, 31, 20, the mean is computed by adding the values—25, 29, 14, 18, 30, 31, 20 = 168—and dividing by the sample size, $n = 7$. Using this data, we get $\bar{x} = \frac{168}{7} = 24$. Our point estimate for the mean of the entire population would be 24.

By the same token, if in a given sample of 50 men who are questioned, we find 22 Republicans, then $\frac{x}{n} = \frac{22}{50} = .44$ is to be the point estimate for p.

This same principle also applies to the point estimate for the variance (σ^2). The estimate, remember, is based on a specific set of values (sample) from a population.

EXAMPLES

1) From the following set of given values, obtain the point estimates for the mean, variance, and standard deviation:

$$14, 11, 9, 10, 5, 6, 12, 16, 7,$$

i) $\bar{x} = \dfrac{14 + 11 + 9 + 10 + 5 + 6 + 12 + 16 + 7}{9}$

$\bar{x} = \dfrac{90}{9} = 10$, the point estimate of μ.

ii) $\hat{s}^2 = \Sigma \dfrac{(x - \bar{x})^2}{n - 1}$ (for ungrouped data)

$= \dfrac{(14 - 10)^2 + (11 - 10)^2 + (9 - 10)^2 + (10 - 10)^2 +}{}$
$\dfrac{(5 - 10)^2 + (6 - 10)^2 + (12 - 10)^2 + (16 - 10)^2 + (7 - 10)^2}{9 - 1}$

$= \dfrac{(4)^2 + (1)^2\ (-1)^2 + (0)^2 + (-5)^2 + (-4)^2 + (2)^2 + (6)^2 + (-3)^2}{8}$

$= \dfrac{16 + 1 + 1 + 0 + 25 + 16 + 4 + 36 + 9}{8}$

$= \dfrac{108}{8} = 13.5.$

However,

iii) $\sigma^2 = \Sigma(x - \bar{x})^2 = \dfrac{108}{9 - 1} = \dfrac{108}{8} = 13.5$, the point estimate for the variance.

$$\sigma = \sqrt{13.5} = 3.674$$

2) When inspector #6 at the Stern Jeans Factory checked 180 pair of jeans one morning, he found 29 to be irregular. What was the point estimate p for the proportion or irregular jeans?

$$\text{The point estimate of } p = \frac{x}{n} = \frac{29}{180} = .161$$

CONFIDENCE INTERVALS

When the purpose of our research is to estimate the unknown mean (μ) of a population based on the mean ($\bar{x}$) of a given corresponding sample, this estimation is enhanced by determining a range of values which we are confident contains the population mean. This range of values is called a *confidence interval*.

The method of interval estimation (confidence intervals) involves the computation of 2 numbers (points) and the construction of an interval within which the parameter lies with a specific degree of confidence.

Interval estimates will be provided for the population parameters μ, σ^2, and p. For the parameter μ, two situations are considered: one, when σ^2 is known, and a second, when σ^2 is unknown and has to be estimated from the sample data.

When we estimate confidence intervals for the population parameter in question, we will determine *confidence limits* for the estimate. For example, if we know that $\bar{x}$ = 100, it is likely that it will be between 90 and 110. 90 will then represent the lower confidence limit and 110 the upper confidence limit. Statisticians might say: "A 95% confidence interval for the parameter is found in the interval (90, 110). It means that the speaker is 95% confident that the parameter μ lies between 90 and 110. The 95% represents a .95 *coefficient of confidence*.

NOTE: A confidence coefficient is different than probability. If there is a .90 coefficient of confidence, it means that if a number of samples were used, 90% of the intervals tested would contain the population parameter sought.

Confidence Intervals for μ

1) When the data is quantitative, the variance (σ^2) is known, and the population is normal:

Step 1: Decide which confidence coefficient to use $(1 - \alpha)$.

Step 2: Draw a sample of size n and compute the value of $\bar{x}$ which is the point estimate of μ.

Step 3: From the known variance σ^2, calculate $\frac{\sigma}{\sqrt{n}}$.

Step 4: Using Appendix Table 2 for normal distributions, find $z_{1 - \alpha/2}$ from the coefficient $(1 - \alpha)$ selected.

NOTE: If the confidence level sought is 95%, then $1 - \alpha = z_{1 - .05/2} = z_{.95/2} = z_{.475}$

This translates to .5 − .025 in Appendix Table 1 (page 119) which is .4750. This corresponds to a Z score of 1.96 in the table.

Step 5: Compute the formula for confidence limits when σ^2 is unknown.

$$\bar{x} - Z_{1 - \alpha/2} \frac{\sigma}{\sqrt{n}} < \mu < \bar{x} + Z_{1 - \alpha/2} \frac{\sigma}{\sqrt{n}}$$

The confidence interval for μ is calculated as the $\bar{x} \pm$ the confidence limits.

Examples

1) The years of service, when a group of 36 policemen were sampled, averaged 15 years with a variance of 9 years. Calculate the interval for μ with a confidence limit of 95%.
   i) From the information we get:
   $$\bar{x} = 15 \quad \sigma^2 = 9 \quad n = 36, \quad \text{and} \quad \alpha = 1 - .95 = .05$$

   If $\sigma^2 = 9$, then $\sigma = \sqrt{9} = 3$
   $$\frac{\sigma}{\sqrt{n}} = \frac{3}{\sqrt{36}} = \frac{3}{6} = .5$$

   ii) If $\alpha = .05$, then (from Appendix Table 2) $Z_{\alpha/2}$ or $Z_{.025}$ corresponds to .4750, which makes $Z_{1-\alpha/2} = 1.96$.
   iii) Using the formula, we get:
   $$\bar{x} - Z_{1-\alpha/2} \frac{\sigma}{\sqrt{n}} < \mu < \bar{x} + Z_{1-\alpha/2} \frac{\sigma}{\sqrt{n}} =$$
   $$15 - 1.96(.5) < \mu < 15 + 1.96(.5) =$$
   $$15 - .98 < \mu < 15 + .98 =$$
   $$14.02 < \mu < 15.98$$

   The confidence limits for a .95 coefficient of confidence are (14.02, 15.98).

2) Using the data in Example 1), find the confidence interval for a .98 coefficient of confidence.
   i) $\bar{x} = 15 \quad \sigma^2 = 9 \quad n = 36 \quad \alpha = 1 - .98 = .02$

   ii) $\frac{\sigma}{\sqrt{n}} = .5 \, Z_{1-\alpha/2} = 2.33$ (.4901 is closest value to .4900)

   iii) $\bar{x} - Z_{1-\alpha/2} \frac{\sigma}{\sqrt{n}} < \mu < x + Z_{1-\alpha/2} \frac{\sigma}{\sqrt{n}} =$
   $$15 - 2.33(.5) < \mu < 15 + 2.33(.5) =$$
   $$15 - 1.165 < \mu < 15 + 1.165 =$$
   $$13.835 < \mu < 16.165$$

   The 98% confidence limits are (13.835, 16.165)

3) When the data is quantitative, the variance (σ^2) is *not* known and the population is normal:
   Step 1: Determine the sample size n.
   Step 2: Decide upon the coefficient of confidence to be tested $(1 - \alpha)$ and find $\tau_{\alpha/2}$ $(n - 1) df$ in the τ-distribution table.
   Step 3: Compute the sample mean $(\bar{x})$, compare the sample standard deviation (s) and then calculate $s/\sqrt{n}$ for the sample data.
   Step 4: Solve for the formula for confidence limits when σ^2 is *un-known*.

$$\bar{x} - \tau_{\alpha/2} \frac{s}{\sqrt{n}} < \mu < \bar{x} - \tau_{\alpha/2} \frac{s}{\sqrt{n}}$$

Examples

1) From a normal population, we have a sample of 25 women with a mean age of 28 and a variance (s^2) of 72 years. What is the interval for a .95 coefficient of confidence of the population mean?

   i) $n = 25$

      $\bar{x} = 28$

      $s = \sqrt{72} = 8.49$

      $\dfrac{s}{\sqrt{n}} = \dfrac{8.49}{\sqrt{25}} = \dfrac{8.49}{5} = 1.698$

   ii)   for $n - 1\ df, \tau_{\alpha/2} = \tau_{24,.025}$

NOTE: $1 - \alpha = 1 - .95 = .05 \ .\alpha/2 = .025$

From Appendix Table 2 (for τ-distributions (page 126), we get $\tau_{24,.025}$ = 2.064

   iii)   Using the formula, we have:

$$\bar{x} - \tau_{\alpha/2} \frac{s}{\sqrt{n}} < \mu < \bar{x} + \tau_{\alpha/2} \frac{s}{\sqrt{n}} =$$

      $28 - 2.064(1.698) < \mu < 28 + 2.064(1.698) =$

      $28 - 3.505 < \mu < 28 + 3.505$

      $24.495 < \mu < 31.505$

   iv)   The 95% confidence interval for a μ of 28 is (24.495, 31.505)

2) Using the same data in Example 1), what is the 98% confidence interval?

   i) $n = 25$

      $\bar{x} = 28$

      $s = \sqrt{72} = 8.49$

      $\dfrac{s}{\sqrt{n}} = \dfrac{8.49}{\sqrt{25}} = \dfrac{8.49}{5} = 1.698$

   ii)   For $n - 1\ df, \tau_{\alpha/2} = \tau_{24,.01} = 2.492$

   iii)   Using the formula, we have:

$$\bar{x} - \tau_{\alpha/2} \frac{s}{\sqrt{n}} < \mu < \bar{x} + \tau_{\alpha/2} \frac{s}{\sqrt{n}}$$

      $28 - 2.492\,(1.698) < \mu < 28 + 2.492\,(1.698)$

      $28 - 4.23 \qquad\quad < \mu < 28 + 4.23$

      $23.77 \qquad\qquad\ < \eta < 32.23$

   iv)   The 98% confidence interval is (23.77, 32.23)

Confidence Intervals for σ^2

When the data is quantitative and the population is normal, the confidence interval for the variance (σ^2) is computed using the formula:

$$\left(\frac{(n-1)\,s^2}{x^2 n - 1,\,\alpha/2} < \sigma^2 < \frac{(n-1)\,s^2}{x^2 n - 1,\,1 - \alpha/2} \right)$$

where,

n = sample size
$n - 1$ = the number of degrees of freedom
$\hat{s}^2$ = the sample variance

Examples

1) In Colorado where the snowfall for the winter is normally distributed, the mean snowfall (annually) is 120 inches. From a sample of 16 years we get a variance (s^2) of 20 inches. What is the confidence interval for σ^2 with a coefficient of confidence of .95?

  i) $\bar{x} = 120$
       $n = 16$
       $\hat{s}^2 = 20$

  ii) Using the chi-square (x^2) distribution, Appendix Table 4 (page 128), for an area α of .05, we get:

$$x^2_n - 1,\,\alpha/2 = x^2_{15,\,0.025} = 27.488$$

and

$$x^2_n - 1,\,1 - \alpha/2 = x^2_{15,\,0.975} = 6.262$$

  iii) Employing the formula to compute the confidence interval for σ^2, we get:

$$\frac{(n-1)\,s^2}{x^2_{n-1,\alpha/2}} < \sigma^2 < \frac{(n-1)\,s^2}{x^2_{n-1,1-\alpha/2}} =$$

$$\frac{(16-1)\,20}{27.488} < \sigma^2 < \frac{(16-1)\,20}{6.262} =$$

$$\frac{300}{27.488} < \sigma^2 < \frac{300}{6.262}$$

$$10.914 < \sigma^2 < 47.908$$

The 95% confidence interval for σ^2 is (10.914, 47,908)

2) On the Santa Monica Freeway one Friday afternoon, 10 cars picked at random averaged 58 miles per hour with a variance of 28. What is the confidence interval for a 90% coefficient of confidence for this normal population?

  i) $\bar{x} = 58$
       $n = 10$
       $\hat{s}^2 = 28$

  ii) Using Appendix Table 4 (page 128), for x^2-distributions, for an area (α) of .10, we get:

$$x^2_{n-1, \alpha/2} = x^2_{9, .05} = 16.919$$
$$x^2_{n-1, 1-\alpha/2} = x^2_{9, 0.95} = 3.325$$

iii)   Employing the formula, we get:

$$\frac{(n-1) s^2}{x^2_{n-1, \alpha/2}} < \sigma^2 < \frac{(n-1) s^2}{x^2_{n-1, 1-\alpha/2}}$$

$$\frac{(10-1) \, 28}{16.919} < \sigma^2 < \frac{(10-1) \, 28}{3.325} =$$

$$\frac{252}{16.919} < \sigma^2 < \frac{252}{3.325} =$$

$$14.894 < \sigma^2 < 75.789$$

iv)   The confidence interval for a .90 coefficient of confidence for σ^2 is (14.894, 75.789)

Confidence Intervals for p

For One Population

When the data is *qualitative* and we are testing a binomial distribution for p, the probability of success, the formula employed is:

$$\frac{x}{n} - Z_{1-\alpha/2} \sqrt{\frac{\frac{x}{n}(1 - \frac{x}{n})}{n}} < p < \frac{x}{n} + Z_{1-\alpha/2} \sqrt{\frac{\frac{x}{n}(1 - \frac{x}{n})}{n}}$$

where,

n = the sample size
x = the number of successes

NOTE: To solve problems involving the confidence limits for p, we use the normal distribution (z) table to compute $z_{1-\alpha/2}$.

Examples

1)   When a sample of 100 randomly selected boys took a taste test comparing 2 cereals, 62 said they favored "Sugar Crispies" over "Animal Dunkers." What is the confidence interval for p, the proportion of the boys choosing "Sugar Crispies," based on a 98% coefficient of confidence?

i) $x = 62$
$n = 100$
$\frac{x}{n} = \frac{62}{100} = 0.62$

ii) $Z_{1-\alpha/2} = Z_{1-.02/2} = Z_{.98/2} = Z_{.4900}$
From the normal curve, Appendix Table 2, we get:

$$.4901 = 2.33$$

iii) Because $\sqrt{\dfrac{\dfrac{x}{n}(1-\dfrac{x}{n})}{n}}$ occurs on both sides of the inequality, it should be computed once and inserted on both sides of p.

$$\sqrt{\dfrac{\dfrac{x}{n}(1-\dfrac{x}{n})}{n}} = \sqrt{\dfrac{.62(1-.62)}{100}}$$

$$= \sqrt{\dfrac{.62(.38)}{100}}$$

$$= \sqrt{\dfrac{.2356}{100}}$$

$$= \sqrt{.0023}$$

$$= .0485$$

iv) Now, substituting in the formula, we have:

$$\dfrac{x}{n} - Z_{1-\alpha/2}\sqrt{\dfrac{\dfrac{x}{n}(1-\dfrac{x}{n})}{n}} < p < \dfrac{x}{n} + Z_{1-\alpha/2}\sqrt{\dfrac{\dfrac{x}{n}(1-\dfrac{x}{n})}{n}}$$

$$.62 - 2.33(.0485) \quad < p < .62 + 2.33(0.485) =$$
$$.62 - .113 \quad\quad < p < .62 + .113$$
$$.507 \quad\quad\quad < p < .733$$

The 98% confidence interval is (.507, .733)

2) When a flu vaccine was given to 80 senior citizens, 52 of them did not get the flu in 1981. What is the confidence interval for p, the proportion of healthy senior citizens, based on a .95 coefficient of confidence?

i) $x = 52$
$n = 80$
$\dfrac{x}{n} = \dfrac{52}{80} = 0.65$

ii) $Z_{\alpha/2} = Z_{1-.05/2} = Z_{.95/2} = Z_{.475}$
From the z table, .4750 corresponds to 1.96.

iii) $\sqrt{\dfrac{\dfrac{x}{n}(1-\dfrac{x}{n})}{n}} = \sqrt{\dfrac{.65(.35)}{80}}$

$$= \sqrt{\dfrac{.2275}{80}}$$

$$= \sqrt{.0028}$$

$$= 0.0533$$

iv) $\frac{x}{n} - Z_{1-\alpha/2} \sqrt{\frac{\frac{x}{n}(1-\frac{x}{n})}{n}} < p < \frac{x}{n} + Z_{1-\alpha/2} \sqrt{\frac{\frac{x}{n}(1-\frac{x}{n})}{n}} =$

$.65 - 1.96 \,(.0533)$ $< p < .65 + 1.96 \,(.0533) =$

$.65 - .0144$ $< p < .65 + .1045$

$.5455$ $< p < .7545$

The 95% confidence interval is (.5455, .7545).

For 2 Populations

When the data is qualitative and where p_1 is the proportion for population 1 and p_2 is the proportion for population 2, we compute the confidence interval for $p_1 - p_2$ from 2 samples using the following formula:

$$\left(\frac{x}{m} - \frac{y}{n}\right) - Z_{1-\alpha/2} \sqrt{\frac{\frac{x}{m}(1-\frac{x}{m})}{m} + \frac{\frac{y}{n}(1-\frac{y}{n})}{n}} < p_1 - p_2 < \left(\frac{x}{m} - \frac{y}{n}\right) +$$

$$Z_{1-\alpha/2} \sqrt{\frac{\frac{x}{m}(1-\frac{x}{m})}{m} + \frac{\frac{y}{n}(1-\frac{y}{n})}{n}}$$

where,

         x = number of successes in population 1
         m = sample size of population 1
         y = number of successes in population 2
         n = sample size of population 2

Example

Of 100 parents interviewed in Los Angeles, 65 favored raising the driving age to 18. In San Francisco, 59 out of 110 favored raising the legal age to drive. What is the confidence interval for a coefficient limit of 95% for the difference between p_1 and p_2?

i) p_1 = Los Angeles p_2 = San Francisco
     x = 65 y = 59
     m = 100 n = 110
     $\frac{x}{m} = \frac{65}{100} = 0.65$ $\frac{y}{n} = \frac{59}{110} = 0.536$

ii) $\frac{x}{m} - \frac{y}{n} = 0.65 - 0.536 = 0.114$

iii) $Z_{\alpha/2} = Z_{.05/2} = 5 - .025 = .4750$
From Appendix Table 1 (page 119), $.4750 = 1.96$

iv) $\sqrt{\frac{\frac{x}{m}(1-\frac{x}{m})}{m} + \frac{\frac{y}{n}(1-\frac{y}{n})}{n}} = \sqrt{\frac{.65\,(1-.65)}{100} + \frac{.536\,(1-.536)}{110}}$

$$= \sqrt{\frac{.65\,(.35)}{100} + \frac{.536\,(.464)}{110}}$$

$$= \sqrt{\frac{.2275}{100} + \frac{.2487}{110}}$$
$$= \sqrt{.0023 + .0023}$$
$$= \sqrt{.0046}$$
$$= .068$$

v) Substituting in the formula for the confidence interval for 2 proportions, we get:

$0.114 - 1.96 \, (0.068) < p_1 - p_2 < 0.114 + 1.96 \, (0.068) =$
$0.114 - 0.129 \quad\quad < p_1 - p_2 < 0.114 + 0.133 =$
$0.019 \quad\quad\quad\quad < p_1 - p_2 < 0.247$

The 95% confidence interval for the difference $p_1 - p_2$ is (−0.019, 0.247).

EXERCISES

Exercise 7-1: Nine randomly selected students who took a math quiz got scores of: 65, 79, 83, 59, 87, 91, 70, 67, 81. What are the point estimates for the following:

a) the mean
b) the variance
c) the standard deviation

Exercise 7-2: The following temperature readings were obtained in New York during the month of October: 59, 66, 48, 71, 77, 54, 60, 49, 50, 66, 58, 63, 70, 51, 72, 67. Assuming that the temperature was normally distributed, determine the confidence intervals for the population mean (μ), using the following coefficients of confidence:

a) .98
b) .90
c) .95

Exercise 7-3: Using the data in Exercise (7-2), determine the confidence intervals for the variance (σ^2) using the following coefficients of confidence:

a) .90
b) .95
c) .98

Exercise 7-4: When a drug was tested on laboratory mice with a virus, 70 of the 120 mice tested survived. Determine the confidence intervals for p, using the following coefficients of confidence:

a) .90

b) .92
c) .88

Exercise 7-5: 75 women and 80 men were polled on their opinion about gun control. 55 of the women and 48 of the men favored it. Determine the confidence interval for $p1 - p_2$ (p_1 = women, p_2 = men) for a .98 coefficient of confidence.

8

HYPOTHESIS TESTING

"Making Decisions About Population Characteristics"
or "To Err May Be Human, But Is Poor Statistics"

In this chapter, we will use the collected and analyzed (measures of location) data to make statistically based decisions in our empirical world. A hypothesis is an educated guess about the nature of the parameters in a specific population. Hypothesis testing forms the basis of inferential statistics, in which conclusions (inferences) about the nature of a population are made on the basis of data (observations) drawn from the population.

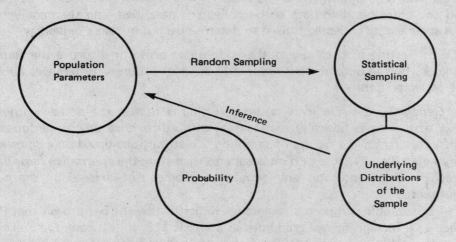

Figure 8A. Inferential reasoning.

INFERENTIAL STATISTICS

We start with a population and determine the decisions we would like to make about the quantitative and qualitative parameters. Next, we select a random sample and calculate the statistical measures of the sample ($\bar{x}$, s^2, s, p). These statistics reflect corresponding parameters in the population. From observing the sample statistics, we can infer back to the parameters in light of the underlying distributions (normal, τ-student, chi-square and f) and probability theory (binomial distributions).

TESTING OF THE HYPOTHESIS FOR THE MEAN: THE MECHANICS OF HYPOTHESIS TESTING

In a step-by-step process, we will determine whether a hypothesis concerning a parameter (i.e., μ) is justifiable and therefore useful. If our testing indicates the hypothesized value is tenable, we retain the hypothesis; if it is untenable, we then reject the hypothesis.

State the Hypothesis to Be Tested

In the application of statistics to research in a myriad of disciplines, a hypothesis provides the general scope of the investigation, and works to delineate the statement of the specific problem and the variables we wish to investigate. The hypothesis is the basis for the research methods to be used and the corresponding data to be collected, described, and then analyzed. The sample data collected is used to determine a hypothesis's tenability.

NOTE: Testing a specific hypothesis does *not* prove or disprove the stated conjecture. The outcome only works to support a theory (inference) about the collected data.

Generally, the rule in hypothesis testing is to test the *null hypothesis* (H_0) first. This is basically a statement of no difference or no relationship when comparing the results of sampling 2 distributions based on a common parameter (i.e., $\bar{x}$). It is also necessary to then state the *alternative hypothesis* (H_a), which indicates any possible outcomes not covered by the null hypothesis.

For example, suppose a researcher wants to test the hypothesis that the mean I.Q. of high school graduates in a city is 115. In this case, the null hypothesis is that the mean I.Q. is equal to 115, which is written:

$$H_0 : \mu = 115$$

This can also be symbolized by:

$$H_0 : \mu - 115 = 0$$

where,

$$H_0 = \text{the null hypothesis symbol}$$
$$\mu = \text{the population mean}$$
$$115 = \text{hypothesized value to be tested}$$

To state the *alternative hypothesis*, which would include all of the possible outcomes not covered in the null hypothesis, we write:

$$H_a : \mu \neq 115$$

where,

H_a = the alternative hypothesis symbol
$\neq$ = does not equal
115 = hypothesized value to be tested

In this first step of hypothesis testing, the researcher in our example will have to test the null hypothesis that the mean I.Q. equals 115 against the alternative hypothesis that the mean I.Q. does *not* equal 115.

Computing the Test Statistic

Based on our I.Q. example, the second step in the process of testing the hypothesis is to analyze the data collected from the sample. When computing the test statistic (i.e., $\bar{x}$) of our sample data, it is necessary to calculate the *z score*. The value of the test statistic is the basis on which we decide whether to retain or reject our null hypothesis.

For the purpose of our example, let's say the researcher gives an I.Q. test to 100 recent high school graduates and finds the mean ($\bar{x}$) I.Q. to be 112.5 with a standard deviation (σ) of 24 and therefore, a standard error ($\sigma_{\bar{x}}$) that equals $\sigma/\sqrt{n}$ and computes to:

$$\frac{\sigma}{\sqrt{n}} = \frac{24}{\sqrt{100}} = \frac{24}{10} = 2.4$$

Using the same method employed in Chapter 6 in the section, "The Normal Distribution," he then compares the sample mean ($\bar{x}$) to the hypothesized mean (μ) and with the concept of z scores to determine the number of standard errors $\bar{x}$ is from μ, the researcher calculates the test statistic. For this example:

$$\text{Test statistic} = \frac{\text{statistic} - \text{parameter}}{\text{standard error of the sample}}$$

$$z = \frac{\bar{x} - \mu}{\sigma_{\bar{x}}}$$

$$z = \frac{112.5 - 115}{2.4} = \frac{-2.5}{2.4}$$

$$z = -1.04$$

Our sample is, therefore, 1.04 standard errors below the mean (μ) of 115. From Appendix Table 1 (page 119), we calculate the probability of deriving a sample mean ≤ 112.5 is equal to .5 − .3508 or .1495 when the null hypothesis is true. We can conclude that such an occurrence is likely if the null hypothesis is true. 1.04 standard errors is generally not too great to reject the null hypothesis.

NOTE: If the variable (σ^2) is unknown, we must use the τ-student distribution (Appendix Table 3, page 127) to compute the test statistic.

Determining Standards for the Rejection of H₀

In this section, you will be introduced to strategy for determining if a null hypothesis (H_0) should be rejected. The *first step* is for the researcher (that's you) to determine a critical region of the test on both sides of the mean (μ). This area (α) is up to the researcher's discretion. Use $\pm Z_{\alpha/2}$ as the critical region. Thus, you must reject H_0 if:

$$Z \geq Z_{1-\alpha/2}$$
$$\text{or}$$
$$Z \leq -Z_{1-\alpha/2}$$

We accept H_0 otherwise.

Example

A group of concerned senior citizens claims the average rent for an apartment in Los Angeles is $420 per month. From a random sample of 90 apartments, the mean is found to be $410 per month with a standard deviation of $9 per month. What is the validity of the senior citizens' assertion where $\alpha = .10$?

i) The null hypothesis is $H_0 : \mu = 420$.
    The alternative hypothesis is $H_0 : \mu \neq 420$.

ii) The test statistic $Z = \dfrac{\bar{x} - \mu}{\sigma_{\bar{x}}}$

where,

$$\sigma_{\bar{x}} = \frac{\sigma}{\sqrt{n}} = \frac{9}{\sqrt{90}} = \frac{9}{9.487} = .95$$

$$Z = \frac{410 - 420}{.9} = \frac{-10}{9} = -11.11$$

iii) Since

$$\alpha = .10, Z_{1-\alpha/2} = Z_{.90/2} = Z_{.45}$$

From Appendix Table 2 (page 126). we get 1.645 corresponds to .4500 (half-way between 1.64 and 1.65).

$$Z_{.05} = 1.645$$
$$-Z_{.05} = -1.645$$

iv) We reject the null hypothesis if:

$$Z \geq Z_{1-\alpha/2} \geq 1.645$$
$$Z \leq Z_{1-\alpha/2} \leq -1.645$$

$Z = -11.11$. This Is less than -1.645. Hence, we reject the null hypothesis.

The *second step* in establishing criteria for accepting or rejecting H_0 is to be aware of possible errors inherent in deciding if the null hypothesis is tenable. The following two types of error are possible:

a) *Type I Error*—Rejection of the null hypothesis when it is actually true.
b) *Type II Error*—Retention of a null hypothesis that is false.

NOTE: The strength of a test or critical region is assessed by the probabilities of Type I and Type II Errors.

$$\alpha = \text{Probability (reject } H_0/H_0 \text{ is true)} = P \text{ (Type I Error)}$$
$$\beta = \text{Probability (accept } H_0/H_0 \text{ is false)} = P \text{ (Type II Error)}$$

Naturally, the statistician wants α and β to be small. The probability of a Type I Error is called the *size* of the test. $1 - \beta$ (the probability of Type II Error) is called the *power* of the test.

Actual Situation	Inferences	
	H_0 Is True	H_0 Is False
H_0 Is True	No Error: Probability $= 1 - \alpha$	Type I Error: Probability $=$ Size $= \alpha$
H_0 Is False	Type II Error: Probability $= \beta$	No Error: Probability $=$ Power $= 1 - \beta$

The general theory of hypothesis testing concerns balancing the level, or the probability of a Type I Error against the power, the probability of a Type II Error.

It is difficult to say which error has the more dire consequence. Consider the problem of updating (changing the style of) a standardized exam, such as the Scholastic Aptitude Test (SAT). Suppose the hypothesis is made that the new test is just as equitable as the original. If the new test is not as fair, but the hypothesis is rejected (Type I Error), the new exam will be substituted which might prejudice the chances of certain students entering the college of their choice. On the other hand, if the hypothesis is accepted, although the new test is superior (Type II Error), the opportunity to upgrade an important educational tool might be lost.

This illustration shows that only a value judgment can determine if one error is more serious than the other. The ideal situation is to minimize the possibility of either error occurring. To accomplish this, the *third step* in selecting criteria for rejection is to predetermine a *level of significance* (α) that calculates the probability of making a Type I Error if H_0 is rejected.

The two most frequently used levels of significance are 0.05 and 0.01. These areas represent the total (5% or 1%) of the area in a normal distribution that lies in both tails. Since the null hypothesis involves an area either below or above the mean (μ), only 1/2 of the α (.025 or .005) is needed for the test, because the null hypothesis is tested against the alternative hypothesis (H_1) which bases its conjecture *non*directionally, $H_1 : \mu \neq 115$. From Figure 8-B, the region of rejection, the *critical region* below and above the mean is ± 1.96.

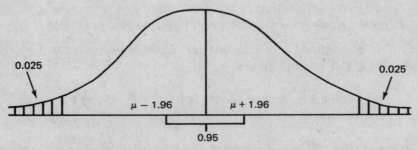

Figure 8-B. A normal distribution used to test H_0 for a 5% (.05) level of significance.

We have been discussing null hypothesis where rejection or acceptance of the alternative hypothesis was nondirectional. That is, we rejected H_0 if the test statistic fell out of the critical region.

$$H_0 : \mu = 115$$
$$H_a : \mu \neq 115 \text{ means } \quad \mu < 115 \text{ or } \mu > 115$$

However, for more critical investigation, testing the null hypothesis against the alternative hypothesis in only one direction is appropriate. From our previous example, the researcher might want to see if the I.Q. level is less than 115. We get:

$$H_0 : \mu = 115$$
$$H_a : \mu < 115$$

NOTE: When testing the alternative hypothesis in one direction, the full level of significance (.05 or .01) should be used in the one-tail test.

Example

A sample of size 36 from a normal population has a standard deviation of 14 and a mean of 44. Test for $H_0 : \mu = 47$ and $H_\alpha : \mu < 47$ for a level of significance of .05.

i) $H_0 : \mu = 47$
   $H_\alpha : \mu < 47$
   $\alpha = .05$

ii) $\sigma_{\bar{x}} = \dfrac{\sigma}{\sqrt{n}} = \dfrac{14}{36} = \dfrac{14}{6} = 2.33$

   $Z = \dfrac{\bar{x} - \mu}{\sigma_{\bar{x}}}$

   $= \dfrac{44 - 47}{2.33}$

   $= \dfrac{-3}{2.33} = -1.288$

iii) $Z_{1-\alpha} = Z_{1-.05} = 1.645$

iv) $1.288 < 1.645$

The critical region is $47 - 1.65$. Because the test statistic falls into that region, we thus accept the null hypothesis and reject the alternative hypothesis.

Deciding to Reject or Retain the Null Hypothesis

By accepting the null hypothesis in the preceding example, the researcher says the value 44, the sample mean, is tenable. If the hypothesis had been rejected, the researcher would be saying that the probability is less than 5% that 44 (the observed sample mean) would have occurred by chance if H_0 was valid.

TESTING OF THE HYPOTHESIS WHEN THE VARIANCE (σ^2) IS UNKNOWN

In testing the null hypothesis, when the variance (and therefore, σ) was known, the normal distribution was used. Z, the test statistic was computed using Appendix Table 2 (page 126), The sampling distribution of the mean is normally distributed and is assumed to be equal to the hypothesized value of the mean with the standard error of the mean (standard deviation) equaling $\sigma/\sqrt{n}$.

However, when σ^2 is unknown, it becomes necessary to use the standard deviation of the sample (s) to estimate σ. Then, $s/\sqrt{n}$ is the estimate of the standard error of the mean, giving us:

$$S_{\bar{x}} = S/\sqrt{n}$$

In hypothesis testing, it is rare when σ^2 is known and, thus, the normal distribution is inapplicable. It becomes necessary to use the Student's τ-distribution. The test statistic is symbolized by:

$$\tau = \dfrac{\bar{x} - \mu}{S_{\bar{x}}}$$

The τ-distribution, as discussed in Chapter 7 is dependent on a population parameter known as the degrees of freedom (df).

NOTE: In hypothesis testing the number of degrees of freedom equals $n - 1$, the sample size minus one.

Testing H_0: $\mu = \mu^0$ Against $H\alpha$: $\mu = \mu_0$, Where μ_0 Is the Hypothesized Mean and σ^2 Is Unknown

1) State the null hypothesis and the alternative hypothesis.
2) Draw a sample size n from a normal population.
3) Compute the test statistic.
4) Determine the level of significance (α level) and compute $\tau_{\alpha/2, n-1}$ for the 2-tailed, nondirectional alternative hypothesis.
5) Decide whether to retain or reject the null hypothesis.

Example

From the following sample data of test scores, test the hypothesis that $\mu = 75$ with a .05 level of significance. 65, 74, 81, 80, 85, 78, 69, 77, 75.

i) $H_0 : \mu = 75$
      $H_a : \mu \neq 75$

ii) $n = 9$

iii) $\bar{x} = \dfrac{684}{9} = 76$

$$\sigma^2 = \frac{(65 - 76)^2 + (74 - 76)^2 + (81 - 76)^2 + (80 - 76)^2 + (85 - 76)^2 + (78 - 76)^2 + (69 - 76)^2 + (77 - 76)^2 + (75 - 76)^2}{9 - 1}$$

$$\sigma^2 = \frac{(-11)^2 + (-2)^2 + (5)^2 + (4)^2 + (9)^2 + (2)^2 + (-7)^2 + (1)^2 + (-1)^2}{8}$$

$$\sigma^2 = \frac{121 + 4 + 25 + 16 + 81 + 4 + 49 + 1 + 1}{8}$$

$$\sigma^2 = \frac{302}{8} = 37.75$$

$$S = \sqrt{37.75} = 6.14$$

$$S_{\bar{x}} = \frac{S}{\sqrt{n}} = \frac{6.14}{\sqrt{9}}$$

$$S_{\bar{x}} = \frac{6.14}{3}$$

$$S_{\bar{x}} = 2.05$$

$$\tau = \frac{\bar{x} - \mu}{S_{\bar{x}}} = \frac{76 - 75}{2.05}$$

$$= \frac{1}{2.05}$$

$$= .49$$

iv) $\tau_{\alpha/2, n-1} = \tau_{.05/2, 9-1} = \tau_{.025, 8}$

From Appendix Table 3 (page 127), we get:

$\tau_{.025, 8} = 2.306$

$-\tau_{.025, 8} = -2.306$

v) $.518 < 2.306$

The test statistic fell within the critical range 75 ± 2.306. We accept the hypothesis that $H_0 : \mu = 75$.

Testing H_0: $\mu = \mu_0$ Against H_a: $\mu < \mu_0$ or H_a: $\mu > \mu_0$ When σ^2 Is Unknown

The procedure for determining whether to reject or accept the proposed hypothesis is the same as when $H_a : \mu \neq \mu_0$, except we reject the hypothesis if:

$$\tau \leq -\tau_{\alpha, (n-1)} \text{ or}$$
$$\tau \geq \tau_{\alpha, (n-1)}$$

Because the alternative hypothesis is directional (either greater or less than the hypothesized value), the level of significance corresponds to only one tail of the Student's τ-distribution.

Example

Using the data in the preceding example, test the hypothesis that $\mu = 75$ with a .05 level of significance against the alternative hypothesis than H_a : $\mu > 75$.

i) $H_0 : \mu = 75$

$H_a : \mu > 75$

ii) $n = 9$

iii) $\bar{x} = 76$

$S_{\bar{x}} = 1.93$

$\tau = .518$

iv) $\tau_{\alpha, (n-1)} = \tau_{.05, 8} = 1.860$

v) $.518 < 1.860$

We accept the hypothesis $H_0 : \mu = 75$.

ADDITIONAL TESTS OF HYPOTHESIS

There are many research applications of statistical analysis that are not concerned with population means. Consider the following:

a) What is the proportion of women college graduates attending post-graduate programs?

b)   What is the relationship between race and place of origin in determining educational success?

c)   Does the level of success vary for nonwhite female students?

These questions, 3 among thousands, cannot be answered using the statistic, the sample mean ($\bar{x}$), as the basis for comparison. In this section, we will discuss hypothesis testing that relates to population values of proportions (P) and variances (σ^2).

Hypothesis Testing About Proportions

A proportion is the fractional part (number of successes/sample size) of a population that possesses a certain characteristic. From a group of 60 Republican candidates, 27 are elected to office. The proportion of elected Republicans is therefore, 27/60 or 0.45.

NOTE: We will use P to symbolize a proportion. The Greek letter X (Pi) you might see in other sources is too easily confused with π, the math constant used in geometry formulas.

The Hypothesis Test of One Proportion: $H_0 : P = \alpha$

$0 \leq P \leq 1$. P can take a value between 0 and 1 and we can formulate and test any hypothesis that falls between 0 and 1. For purpose of discussion, let's just suppose that from a random sample of 200 women interviewed, 125 favored men who were smarter than they were. The proportion of women responding favorably becomes 125/200 or .625. In a 5-step procedure, we will test that hypothesis.

i)   *Formulate the Hypothesis:* The null hypothesis states that for the entire population of women

$$H_0 : P = 60$$

The alternative hypothesis (nondirectional and 2-tailed) is

$$H_a : P \neq 60$$

ii)   *Determine the Level of Significance:* .05 or .01 are the most commonly used. For this example, $\alpha = .05$.

iii)   *Compute the Test Statistic:* For our example, the statistic being tested is the sample proportion which was 125/200 or .625. We will use the normal distribution (the sample size of 200 is sufficient) to approximate the binomial distribution. This gives us a sampling distribution with the following traits:

1)   The mean of the sampling distribution equals P, the population proportion.

2)   The standard error of a proportion (standard deviation) is symbolized by

$$\sigma_p = \sqrt{\frac{PQ}{n}}$$

where,

P = population proportion possessing the characteristic in question
Q = 1 − P, the proportion of the population that does *not* possess the characteristic in question

In our example, P and Q for the population are unknown. Therefore, we estimate their values. This gives us:

$$S_p = \sqrt{\frac{PQ}{n}}$$

where,

S_p = the standard error of the sample proportion
p = the sample proportion possessing the characteristic in question
q = the sample proportion that does *not* possess the characteristic in question $q = 1 - p$

For our example,

$$S_p = \sqrt{\frac{(.625)(.375)}{200}} = \sqrt{\frac{.234}{200}}$$
$$= \sqrt{.0012}$$
$$= 0.035$$

Our sampling distribution of the proportion is being approximated to a normal distribution. Therefore, the test statistic Z will be employed in the standard formula for the test statistic.

$$Z = \frac{\text{statistic} - \text{hypothesized parameter}}{\text{standard error of sample}}$$
$$Z = \frac{p - P}{S_p}$$
$$Z = \frac{.625 - .60}{0.035} = \frac{.025}{0.035}$$
$$Z = .735$$

iv) *The Critical Region for Rejecting or Retaining the Null Hypothesis:* The sampling distribution is considered normally distributed. Using Appendix Table 2 (page 126), we calculate the critical value of $Z_{\alpha/2}$ because H_a is *nondirectional,* making this a 2-tailed test.

$$Z_{1-\alpha/2} = Z_{1-.05/2} = Z_{.95/2}$$
$$Z_{.475} = 1.96$$
$$-Z = -1.96$$

v) *The Decision to Retain or Reject the Null Hypothesis:* The test statistic Z we calculated at .735 did not exceed the critical value of 1.96. Therefore, we accept the null hypothesis, $H_0 : P = 60$.

The Hypothesis Test that 2 Proportions are Equal

In this section, we will test the hypothesis that 2 population proportions are equal. The two samples involved in the test will be independent for our purposes. We will concern ourselves with dependent samples in this book.

The example we will follow in this and the next section on hypothesis testing involves cigarette smoking. Consider a researcher who asserts that there is no difference in the proportions of cigarette smokers in the age groups 16-25 and 26-35. To prove his theory, the researcher interviews 2 samples. Out of 300 people in the 16-25 group, 130 (130/300 or .433) said they smoked. From 280 people in the 26-35 group, 115 (115/280 or .411) admitted to being smokers.

i) *Formulate the Hypothesis:* The null hypothesis for our example will state that there is no difference in the proportions of people in both age groups who smoke.

$$H_0 : P_1 - P_2 = 0 \quad \text{or} \quad H_0 : P_1 = P_2$$
$$H_a : P_1 - P_2 \neq 0 \quad \text{or} \quad H_a : P_1 \neq P_2$$

This is a 2-tailed test because H_a is nondirectional. For this test, the level of significance should be .01 (.005 for each tail).

ii) *Calculate the Test Statistic:* The test statistic for this hypothesis is the difference between the 2 sample populations.

NOTE: If the sampling distribution is fairly large (at least 200), it will be normally distributed. The hypothesized value of the difference between the 2 population proportions $(P_1 - P_2)$ is the mean (μ) of the sampling distribution. The standard error of the difference between proportions is the standard deviation which is computed by:

$$S_{p1-p2} = \sqrt{pq\left(\frac{1}{n_1} + \frac{1}{n_2}\right)}$$

where,

p = proportion from the 2 samples with the smoking characteristic or

$$\frac{f_1 + f_2}{n_1 + n_2}$$

q = $1 - p$
f_1 = number of smokers in 1st sample
f_2 = number of smokers in 2nd sample

and computing, we get:

$$p_1 = \frac{130}{300} = 0.433$$
$$p_2 = \frac{115}{280} = 0.411$$

$$p = \frac{f_1 + f_2}{n_1 + n_2} = \frac{130 + 115}{300 + 280} = \frac{245}{580} = 0.422$$

$$q = 1 - p = 1 - .422 = .578$$

$$S_{p1-p2} = \sqrt{(.422)(.578)\left(\frac{1}{300} + \frac{1}{280}\right)}$$

$$= \sqrt{.244\left(\frac{1}{300} + \frac{1}{280}\right)} = \sqrt{.244(.033 + .0036)}$$

$$= \sqrt{.00081 + .00087}$$

$$= \sqrt{.0017}$$

$$= .041$$

The sampling distribution is normal, therefore we use Z as the test statistic:

$$Z = \frac{\text{hypothesized value} - \text{population parameter}}{\text{standard error of the proportions}}$$

$$Z = \frac{(p_1 - p_2) - (P_1 - P_2)}{S_{p1-p2}}$$

$$Z = \frac{(.433 - .411) - 0}{.041}$$

$$= \frac{.022}{.041}$$

$$= .537$$

iii) *Determine the Critical Region for Rejecting or Retaining the Null Hypothesis:* Using Appendix Table 2 (page 126), we get:

$$Z_{1-\alpha/2} = Z_{1-.01/2} = Z_{.99/2} = Z_{.495} = 2.58$$
$$-Z_{1-\alpha/2} = -2.58$$

iv) *The Decision to Reject or Retain the Null Hypothesis:* The test statistic, 0.537, is less than the area, 2.58, in the right tail. We retain (accept) the null hypothesis.

NOTE: When testing the hypothesis that there is a difference between 2 population proportions (i.e., $H_0 : P_1 - P_2 = .05$ and $H_a : P_1 - P_2 > .05$), the difference is the population parameter in the formula for computing the test statistic Z. Otherwise, the steps for testing the hypothesis are the same as any one-tailed directional test.

Hypothesis Testing About Variances

These final tests of hypotheses (at least for now) differ from those for means and proportions in that the chi-square (x^2) — distribution and the f-distribution form the sampling distributions to be tested.

The first test of variances is the one-sample situation, where $H_0 : \sigma^2 = a$. The second test, a 2-sample situation, is symbolized by $\sigma_1^2 = \sigma_2^2$. Both of these hypotheses are tested in the same manner, although different sampling distributions are employed.

The One-Sample σ^2 Hypothesis

When we test a hypothesis about the value of a population variance (σ^2), we use the sample statistic S^2 to estimate. The test statistic used is x^2 and it is distributed with $n - 1$ degrees of freedom.

Let us suppose that a sports researcher believes that the foul shooting accuracy in the NBA was more homogeneous (less dispersion) in 1971 than in 1981 where the variance was 289. To test his theory, the researcher took a sample of 30 players from 1971 and computed the variance (s^2) to be 227.

i) *Formulate the Hypothesis:*
$H_0 : \sigma^2 = 289$
$H_a : \sigma^2 < 289$

For this test, we will use .01 level of significance.

ii) *Calculate the Test Statistic:* The sampling distribution of the sample variance conforms to a chi-square (x^2) distribution, the shape of each determined by the number of degrees of freedom. The test statistic for testing the hypothesis of the variance is:

$$x^2 = \frac{(n - 1)s^2}{a}$$

where,

$n - 1 =$ the number of degrees of freedom
$s^2 \quad = \dfrac{(\bar{x} - x)^2}{n - 1}$
$a \quad =$ the hypothesized value of σ^2

From the data in our example, we get:

$s^2 \quad = \dfrac{(30 - 1)\,(227)}{289}$

$\quad = \dfrac{6583}{289}$

$\quad = 22.78$

iii) *Determine the Critical Region for Rejecting or Retaining the Null Hypothesis:* The x^2-distribution and the appropriate sampling distribution are dependent on the number of degrees of freedom. If the calculated value of the test statistic is found to be outside the table's value for the *df* and the level of significance, the null hypothesis is thereby rejected.

Because the critical region is in the left tail and the level of significance is .01 or (1%), .99 (or 99%) of the area is to the right. From Appendix Table 4 (page 128), for (30 − 1) *df* and 99% of the area to the right, we get:

$$x^2_{29,.99} = 14.256$$

iv) *The Decision to Retain or Reject the Null Hypothesis:* To reject the null hypothesis that $\sigma^2 = 289$, the calculated value of x^2 must be less than 14.256. It is *not*, therefore, we accept (retain) the hypothesis.

The 2-Sample σ^2 Hypothesis

In this hypothesis, we will test whether the variance of 2 populations are equal. To test this hypothesis, we use a different sampling distribution, the f-distribution to determine if the variance of population 1 equals the variance of population 2.

NOTE: The f-distribution is a family of distributions requiring 2 values for the degrees of freedom to identify a specific distribution. The values can range from 0 to ∞ and the distributions are not symmetrical.

The example for this section involves the hypothesis that Democratic Congressmen are more homogeneous than Republicans in their voting records. From the Congressional Record, a random sample of 31 Democrats and 25 Republicans is drawn. The variance $(s_1{}^2)$ of the Democrats is found to be 60 and the variance $(s_2{}^2)$ of the Republicans is determined to be 50.

i) *Formulate the Hypothesis*

$H_0 : \sigma_1{}^2 = \sigma_2{}^2$
$H_a : \sigma_1{}^2 > \sigma_2{}^2$

Our level of significance will be .05.

ii) *Calculate the Test Statistic:* The test statistic we are interested in is the ratio of the 2 variances.

$$F = \frac{s_1{}^2}{s_2{}^2}$$

where,

s_1 = the larger variance
s_2 = the smaller variance

For our example,

$$F = \frac{60}{50} = 1.2$$

iii) *Determine the Critical Region for Rejecting or Retaining the Null Hypothesis:* The 2 degree(s) of freedom for our distribution are: $(n1 - 1)$ and $(n2 - 1)$ or 30 for the Democrats and 24 for the Republicans. Using Appendix Table 5 (page 130), we consider the distribution with 29 and 24 degrees of freedom with a 5% (.05) level of significance.

$$F_{30, 24 ; .05} = 1.89$$

iv) *The Decision to Reject or Retain the Null Hypothesis:* The calculated test statistic, the F ratio, at 1.2 is less than the critical value of 1.89. We accept the null hypothesis because the probability is less than 5% that the ratio of sample variances occurred by chance if $H_0 : \sigma_1^2 - \sigma_2^2$ were valid.

SUMMARY

This chapter has introduced you to additional uses of inferential statistics. From the estimation of parameters, we have moved to the processes of hypothesis testing. From the available data, a hypothesis is formulated and a test sample(s) is drawn. The appropriate test statistic and the critical region (based on level of significance) are then calculated. A decision to retain or reject the hypothesis, based on probability, is the last step and it is based on the size of the difference between the sample statistic and the population parameter.

The final decision is also affected by:

a) the directional (one-tailed) or nondirectional (2-tailed) nature of the alternative hypothesis
b) Type I and Type II Errors in testing
c) different levels of significance
d) different critical regions

The formula generally used to measure the appropriate test statistic is:

$$\text{Test statistic} = \frac{\text{sample statistic} - \text{population parameter}}{\text{standard error of the sample (standard deviation)}}$$

The critical regions are calculated using Tables 2, 3, 4, and 5 in the Appendix (pages 119-135):

Hypothesis	Appendix Table
1) $H_0 : \mu = (\sigma^2$ is known) $H_a : \mu \neq a$	Table 2, normal distribution
2) $H_0 : \mu = a$ (σ^2 is known) $H_a : \mu < a$	Table 2, normal distribution
3) $H_0 : \mu = \mu_0$ (σ^2 is unknown) $H_a : \mu \neq \mu_0$	Table 3, Student's τ-distribution
4) $H_0 : \mu = \mu_0$ (σ^2 is unknown) $H : \mu \neq \mu_0$	Table 3, Student's τ-distribution
5) $H_0 : P = a$ $H_a : P \neq a$	Table 2, normal distribution
6) $H_0 : P_1 - P_2 = 0$ $H_a : P_1 - P_2 \neq 0$	Table 2, normal distribution
7) $H_0 : \sigma^2 = a$ $H_a : \sigma^2 \neq a$	Table 4, chi-square distribution
8) $H_0 : \sigma_1^2 = \sigma_2^2$ $H_a : \sigma_1^2 > \sigma_2^2$	Table 5, f-distribution

EXERCISES

Exercise 8-1: From a normal population, a sample of 64 is drawn. The $\bar{x}$ is 25 and the s^2 is 36. Test the hypotheses:

a) $H_0 : \mu = 16$ } with a .05 level of
   $H_a : \mu \neq 16$ } significance

b) $H_0 : \mu = 16$ } with a .01 level of
   $H_a : \mu < 16$ } significance

Exercise 8-2: Lawyers in Chicago earn an average of $35,000 per year. From a random sample of 70 lawyers, an executive headhunter found the average to be $32,000 with a standard deviation of $6,000.

With $\alpha = .01$:

a) State the hypothesis.
b) Compute the test statistic.
c) Decide whether to reject or retain the null hypothesis.

Exercise 8-3: In 1981, the entering freshman class at Stern Tech had averaged 540 on the English part of the SAT Test. In 1982, the administration took a sample of 25 students applying for admission to see if their SAT scores matched last year's average. From the following scores, test the appropriate null and alternative hypotheses:

560	610	475	510	535	465	520
590	490	555	600	605	520	
525	510	560	615	530	540	
535	600	585	500	510	565	

Exercise 8-4: A recent *Playboy* survey found that 7.5% of 120 male readers had never looked at the centerfold. Test the appropriate hypothesis using a .05 level of significance.

Exercise 8-5: A researcher decides there should be no difference in the divorce rate among male marathon runners and those men who did little or no exercise. From a random sample of 110 marathoners, 48 were divorced. From another random sample of 115 nonexercisers, 51 were divorced. Using a .05 level of significance, test the appropriate hypothesis.

Exercise 8-6: The ABC Company Worldwide found that on its aptitude test given to all new employees there was a standard deviation of 17. When the same test was given to a control group of 30 nonemployees, the standard deviation was 18.6. With a .01 level of significance, test the hypothesis that the variance equals 17.

Exercise 8-7: A researcher hypothesizes that there was no difference in the results when 2 drugs were administered to 2 groups of alcoholic laboratory mice. From a sample of 41 in Group A, the variance was 275. From a sample of 50 in Group B, the variance was 290. Test whether $\sigma_1^2 = \sigma_2^2$ for a .10 level of significance.

9

USES, MISUSES, AND ABUSES OF STATISTICS IN OUR PRACTICAL WORLD

"The Value of Statistics to You" or "You Can Fool Some of the People . . . "

Do 4 out of 5 dentists really recommend sugarless gum for their patients who chew gum?

In 1980, 2 surveys of female opinion brought about divergent and confusing results. The issue in question was abortion. One survey proclaimed that the large majority of American women of childbearing age favored abortion. The second survey claimed that the opposite was true.

How could that be? Two surveys. The same topic. The samples should have been fair, representative, and large enough to accurately project (estimate) the results for the entire female population.

Why? The hypothesis that linked the 2 surveys was biased from the start. Survey number one had asked if the women believed it was a woman's right to decide to abort a pregnancy. On the other hand, survey number two asked if the women believed abortion was morally wrong. So it was possible to believe that a woman has the right to choose, while also believing abortion was morally wrong for you.

The two surveys made for interesting cocktail-party chit-chat, but poor statistics. In statistical investigations and applied research, it is the start and finish that present the biggest difficulties; that is, selecting the data to be collected and at the end, making inferences and drawing conclusions. Like a novice pilot, once safely off the ground, the novice statistician will probably be able to soar aloft on a fairly even keel until it comes time to land his craft. It is at this point the novice's (or the ill-prepared's) deficiencies become most evident. Depending on the researcher's expertise, the collected, described, and analyzed data can be used effectively, misused, or abused with equal effectiveness.

It is not essential for useful statistical research that the researcher be a technically trained and certified statistician. Misused and abused statistics have an aura of being poorly prepared, while a good analysis carries with it the unmistakable signs that can be recognized by even a layman.

Effective statistics are characterized by the following:

a) The questions are posed without ambiguity.
b) The sources of the data are given.
c) Possible causes of error are discussed.
d) Deficiencies are pointed out before conclusions are made or inferences are drawn.
e) The conclusions are stated with caution and they are limited to the scope of the initial question.
f) If the conclusions are numerically based, they are stated in terms of the probability of success, with the standard error indicated.

Misused or abused statistics entail the following:

a) The questions are, by choice or chance, poorly worded and misleading.
b) There is a minimum of supplementary discussion and/or explanation.
c) Conclusions are stated as if they were definite and final.
d) The analysis is characterized by an appearance of an air of ease and simplicity, which either confuses or beguiles the unwary reader. Unlike with other branches of mathematics, ease and simplicity are *not* earmarks of soundness and effectiveness.

I must stress the importance of exploring the source of the data (random versus specified sample) that go into a statistical table, chart or graph, and of choosing the materials with an unjaundiced view to the use(s) to which they are to be subjected. And, because the theories or statistics are tied closely to the theories of random chance (probability), so that it would be impossible or at least improbable to interpret statistics without first understanding chance, a good statistician incorporates the two disciplines.

One of the most rapidly growing advances of the twentieth century, developing alongside high technology, modern warfare, and fast foods, is the expansion of the social science of statistics. Originating some 150 years ago when facts bearing on the state, and hence the term "statistics," were collected, the subject was involved with information regarding births and deaths, imports and exports and economic fluctuations. In an effort to understand and utilize this data which affected the state and its citizens, the study of statistics gradually took shape. It became obvious that statistics was the ideal scientific technique for all those research subjects that involve the study of large masses (populations and samples) of individuals or facts, such as insurance, economics, sociology, political science, and history. The nineteenth-century scientist, who was securely tied to the physical sciences, saw statistics as simply an artificial mathematics tool for handling research situations too complicated or complex for straightforward, fundamental mathematical techniques. Today, however, the twentieth-century scientist must deal with statistics on a daily basis.

Disciplines like astronomy, physics, and chemistry rely on the available statistical methods. Statistics has found its way into all the physical sciences. Those parts of physics that study the electron and the nucleus of the atom are totally statistical in nature. Although we do not know how one will react individually, we do know how large masses of them will react. Without the available statistical techniques to interpret the data, the atomic bomb and its even more powerful successors would have remained theories.

With the inroads it has made into the study of modern warfare, politics, big business, and entertainment, statistics has proven to be invaluable to the vast majority of us in the art of living.

Yet, statistics is at the same time one of the most misunderstood and most maligned of subjects. It is often said that you can prove anything with statistics. The reason that tenet is widespread is that, in the spirit of congeniality or in the heat of an argument, the distinction between sound and unsound statistics is often conveniently dropped. It is unfortunately a sad fact that the maximum number of statistical conclusions and premises that reach you, the public, are not founded in sound statistical methods and are, more often than not, biased and misleading.

Actually, statistics is a highly developed, technically based and technologically up-to-date field which has made use of the resources available in advanced and theoretical mathematics. For without a competence (an expertise is not required) in mathematics, an understanding of the principles and practices of statistics and probability and their applications to research in the empirical world, is highly improbable. It is not necessary to be a professional statistician to make valuable use of those same principles and practices in one's professional and personal lives. To prove this declaration to your complete satisfaction is the aim of this chapter and those preceding it which gave you the necessary skills and background.

SOME TRUTHS, MYTHS, AND FALLACIES OF PROBABILITY

Although it has been rising rapidly to fame, achievement, and, on college campuses, infamy, the theory of statistics stands atop another mathematical discipline. Statistics was preceded by and has drawn much of its appeal and framework from probability, the theory of chance.

Born from problems that arose in connection with casino gambling, probability had achieved a relative maturity among mathematical disciplines when statistics was in its infancy. Statistical theory would not have come very far without the pliant ideas provided by the laws of chance that are essential to the presentation and analysis of a collection of facts, the essence of statistics. To compare probability theory and statistical theory is to compare the view from both ends of a pair of binoculars. Probability enlarges the image (i.e., test or experiment), while statistics shrinks the image to fit a specific collection of facts. It was only logical then that probability came first, because it would not be possible to understand statistics without it. And, without the inclusion of the ideas of chance and random events, statistics would be as dry and anxiety-producing as many a student, businessman, biologist, and teacher consider it to be. Statistical tables can, with the help of probability theory, be the keys to the likelihood of future events, so that those events come to life in our imaginations. Consider a study on the number of men who wear toupees. Then imagine how you might pleasantly pass some time at your next boring meeting or class trying to verify the results of the study.

Gambling

Until two French mathematicians came up with the theory that chance was something that could be handled numerically, the notion that all life was a proverbial "wheel of fortune" symbolized the unknown, the mysterious, and the bizarre elements in men's lives. The idea of chance was equated with fate, a concept shrouded in superstition and religious mumbo-jumbo.

The first scientist to actually try his talented hand at unraveling the mysteries of rolling dice and other gaming possibilities was Galileo Galilei. Galileo thought gambling problems worth studying and actually wrote a brief tract on solving the intricacies of the game of dice. But it was the Chevalier DeMere, a devoted gambler and educated scientist, and his friend, Blaise Pascal, the father of the adding machine, who saw the theory of probability as only the workings of natural (physical) law.

From Galileo, to Pascal, to the German mathematician, Gauss, who developed the "Gauss Law of Errors," to the Italian, James Bernoulli and down to Sir Edmund Hoyle, the gambler owes a great debt to the science of mathematics. The theory of probability, a boon to all social sciences, has revolutionized the notion of games of chance, showing just how misguided and fallacious were the theories held by gamblers. The average gambler, the kind who really built the world's lavish casinos, has never really appreciated the service.

Gamblers, those men and women for whom games of chance are ingrained in their psyches, are remarkable for their ability to disregard what is mathematically sound, and trust what is vague (human nature) and what is unjustifiable (luck). It is the professional and the wealthy proprietors of gaming establishments who put their trust in experience and odds. Their profits are thus assured.

It is possible for the "average" gamblers to win big at a casino or a racetrack on occasion, sometimes often, but the theory of probability will, if they play long enough, make losers of them all.

In the following sections, I will attempt to illustrate just how fickle Lady Luck can be. Learn to trust experience or learn how it is to be poorer for the experience.

Poker

My own personal favorite, poker, involves a large element of expertise and judgment not required for dice, roulette, or other games of pure chance.

However, the required judgment and the accompanying skill are based strongly on probability theory.

A distinctly American card game, poker gained its initial popularity in the Old West of Bat Masterson, Doc Holliday, and Wild Bill Hickock, who died holding aces and eights, now affectionately known as "the dead man's hand." Today, poker is played all over the world and in its almost infinite variety, poker games consist of: 7-card stud; 5-card stud; 5-card draw poker; high-low poker; low-ball; 3-card monte, and those variations initiated in "dealer's choice" games.

Poker can be played for table stakes (pot limit), $5, $10, and $20 limits, nickels and dimes or for matchsticks if that's all you have. The number of players can vary from 2 to 8, and sometimes 9. I prefer 7 players with the stakes dependent on my current bank account and the other players. For what makes poker unique is the human factor which can, for many, obliterate the probability theory present at the table.

Even a beginning poker player should realize how a hand with excellent prospects in a 3-player game will not have those same prospects (probability) with 7 players. But when one watches a beginning (or just inexperienced) poker player, the belief is quickly thrust upon him that even a rudimentary knowledge of the chances and of the probability principles would improve the novice's game and his diminishing stack of chips. The occasional and totally inevitable winning of a big pot only encourages the continuing of unsound methods. Putting good money after bad in a vain attempt to "protect" what has already been invested in the pot is a total departure from the common sense most people display away from the poker table.

As with all probability experiments, the probability of an outcome in poker is the number of favorable outcomes divided by the total number of outcomes. Table 9-A illustrates the probabilities involved in a game of 5-card stud (no draw). The total number of outcomes is 2, 598, 960. This is obtained by using permutations:

$$n \, Pr = 52 \, P5 = 52 \times 51 \times 50 \times 49 \times 48 = 311,875,200$$

Since this takes into account all the hands in which the cards are just in a different sequence (i.e., A, K, Q, J, 10 is the same as 10, J, Q, K, A), we must divide by the permutations for any individual hand dealt, $n \, Pr = 5 \, Pr = 5 \times 4 \times 3 \times 2 \times 1 = 120$. $311,875,200 \div 120 = 2,598,960$ possible outcomes for 5-card stud. The order I have listed the hands, for you non-poker players, is their preferential order for winning the pot (i.e., a flush is better than a straight).

In games of poker where replacement cards may be drawn after the original hands are dealt, the probability of success, Table 9-A, has to be adjusted for the cards drawn and the specific hand sought. A variety of examples is given in Table 9-B. The number of cards drawn is based on good poker sense (no suckers here).

Even when they know the correct odds, a good poker player must also take into account the size of the present bet, the number of players, his position at the table, and the amount of money already in the pot. This is what makes poker unique among games of chance—the chance and the strategy.

Table 9-A. The Probabilities of 5-Card Poker Hands

Hand	Number of Possibilities	Probability
Straight Flush (5 cards of same suit in sequence)	40	.00002
Four of a Kind (e.g., 4 Kings + any card)	624	.00024
Full House (3 of a kind + a pair)	3,744	.00140
Flush (5 cards of the same suit)	5,108	.00200
Straight (5 cards in sequence)	10,200	.00390
Three of a Kind (e.g., 3 Jacks + any 2 cards)	54,912	.02110
2 Pairs (e.g., 2 Jacks + 2 Tens + any card)	123,552	.04750
1 Pair (e.g., 2 Kings + any 3 cards)	1,098,240	.42260
"Nothing" of Value (5 unmatching cards)	1,302,540	.50120
	2,598,960	.99996

Table 9-B. Probabilities for Hands in 5-Card Draw Poker

Original Hand	Replacement Cards	New Hand	Probability	Odds
1 Pair	3	2 Pair	.1600	5.25 to 1
	3	3 of a Kind	.1140	7.7 to 1
	3	Full House	.0102	97 to 1
	3	Four of a Kind	.0028	359 to 1
	3	Any Improved Hand	.2870	2.48 to 1
2 Pairs	1	Full House	.0850	10.8 to 1
3 of a Kind	2	Full House	.0610	15.4 to 1
	2	Four of a Kind	.0430	22.5 to 1
	*2	Any Improved Hand	.1040	8.6 to 1
	1	Full House	.0640	14.7 to 1
	1	Four of a Kind	.0210	46 to 1
	*1	Any Improved Hand	.0850	10.8 to 1
4 Straight (one spot open)	1	Straight	.0850	10.8 to 1
Flush	1	Flush	.191	4.2 to 1
4 Straight Flush	1	Straight Flush	.0210	46 to 1
	1	Straight or Flush	.256	2.9 to 1

*You poker players should notice the difference when deciding to draw 1 or 2 cards with 3 of a kind.

NOTE: Never draw to an inside straight and don't player poker with strangers.

Roulette

The roulette wheel is gambling's most sophisticated wheel of fortune. Fortunes have been wagered on which of the 38 slots the white ball will come to rest. The slots (compartments) consist of thirty-six numbers (1/2 red, 1/2 black) and 0 and 00. The players can only make wagers on the numbers 1-36. 0 and 00 belong to the casino and comprise the greater (5.26%) part of its advantage. In discussing roulette and the probabilities involved, 2 assumptions must be made: The wheel itself must be free of bias; and, the croupier (the person working the wheel) must be above reproach.

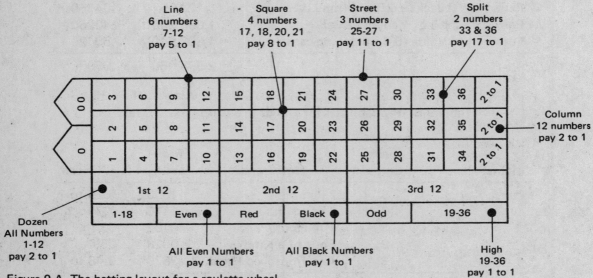

Figure 9-A. The betting layout for a roulette wheel.

Table 9-C. The Probability of the Roulette Wheel (in United States)

Bet	Probability	Correct Odds	Actual U.S. Casino Odds
Straight (any one number 1-36)	1/38 = .0263	37 to 1	35 to 1
Odd or Even (any odd or even number 1-36)	18/38 = .4737*	1.11 to 1	1 to 1
Red or Black (numbers 1-36)	18/38 = .4737	1.11 to 1	1 to 1
Column (12 numbers 1-36)	12/38 = .3157	2.17 to 1	2 to 1
Line (6 numbers 1-36)	6/38 = .1578	5.33 to 1	5 to 1
Square (4 numbers 1-36)	4/38 = .1052	8.5 to 1	8 to 1
Street (3 numbers 1-36)	3/38 = .0789	11.67 to 1	11 to 1
Split (2 numbers 1-36)	2/38 = .0526	18 to 1	17 to 1
Dozen (12 numbers 1-36)	12/38 = .3158	2.17 to 1	2 to 1
High or Low (18 numbers 1-36)	18/38 = .4737	1.11 to 1	1 to 1

*.4737 for odd and .4737 for even add up to .9474. Since the total probability should be 1, that leaves a .0526 (5.26%) advantage for the casino.

NOTE: When discussing odds, we are referring to the ratio of the numerators of the fractions describing probability (not A) to probability (A). When placing a bet on red in roulette, the probability of *not* red is 20/38. The probability of red is 18/38. The odds against red should be 20:18 or 1.11 to 1.

Before we leave the subject of roulette, there is one point I would like to make. If the roulette wheel is functioning without any bias, there is *no* system man can devise that will insure winning at roulette. Machines do not care one whit who is betting or on what their bets are placed. The laws of chance and the theories of probability are indisputable. They can aid you in determining which bets offer you the best odds, but that is all. Casinos love systems players. They will even supply the pencil and paper. It is in their best interest that the rules of roulette and other games agree with the laws of chance.

Lotteries

Lotteries are a simple illustration of the laws of chance. Now legal in many states (i.e., New York and New Jersey), the lottery, as opposed to a raffle, is based on the same principles that helped our country in its infant stages raise $5 million to carry on the Revolutionary War:

1) If for $1 you are given the opportunity for winning one of three prizes—$2,000,000, $100,000 or $50,000—you have 3 chances to win.
2) If 500,00 tickets are sold, you have 3 chances in 500,000 to win a prize.
3) Since $250,000 will be made by the promoters, the probability of success becomes $250,000/$500,000. Looked at as a wager, the odds against the ticketholders are 2 to 1.

However, despite the most unfavorable odds and the weight of the laws of chance on the closing side, the large prizes and the relatively small price for the ticket make lotteries attractive for millions.

An illegal and abused form of the lottery is the "numbers" game found in most cities. Like the legal form, Keno, found in casinos, the "number(s)" is selected by the player. If winning is based on the last 3 numbers of the daily stock market sales, the probability of success should be 1/1000, because there are 1000 possible numbers from 000 to 999. Because the illegal operator or the casino might pay only 600 to 1, you will, based on the laws of chance, lose $400 for every 1000 $1 bets.

Chance Versus Speculation

In the highly uncertain worlds of business and finance, the future is not determined solely from the Laws of Chance. The risks, including bankruptcy, are constantly changing. Even the so-called economic experts are at odds when trying to predict success for any new business or investment. Extraneous risks that do not come into play in probability theory can easily turn chance to speculation. In business, losses from natural disasters, theft, changes in supply and demand, labor problems and political uncertainties worldwide make estimating the probability of success no more than an educated guess.

When comparing the speculator to the participant in a game of chance, we must take into account the skill and the judgment of the speculator. Although the results in a game of craps will be the same no matter who throws the dice, the results of a speculative venture may be dependent on the knowledge of the investor. There are similarities and differences. Let us consider the speculator who invests in the stock market:

a) The stockbroker corresponds to the "banker" of a game of chance who charges an amount proportional to the stakes of the game involved. The commissions paid to the broker correspond to the casino's advantage.

b) As in a game of chance such as baccarat or poker, the speculator is betting against other players (investors) who sell when he buys and vice versa.

c) The total winnings of those players (speculators) who finish with a profit must equal the total losses of the remaining players.

d) The speculator can receive dividends from the companies in which he chooses to invest. Free drinks at a casino are not the same thing.

e) A speculator can invest in both the success (buying stocks) and failure (selling short) of a company whose stock is trading.

f) The "gambler's fallacy" that a long run of either good or bad luck will affect the outcome of the next bet can actually work in speculation. The "principle of momentum" that a market on the upswing tends to continue upward until new economic forces come into play has no parallel in the laws of chance, no matter how many gamblers preach the "streak theory."

g) When the game (i.e., poker) involves both chance and expertise, the gambler is then a speculator. Over the long haul, the relative skills of the participants will determine the winners and losers. The speculator uses his knowledge of a company's profit, its management, its growth rate, its debt structure and its prospects, as well as his knowledge of the stock market in general. The poker player employs his knowledge of human nature, card sense, and the laws of chance which can be calculated.

h) The results of speculation over a short period of time are not a reliable index of the odds for or against any individual speculator. The laws of chance are always reliable indicators for success in gambling.

In conclusion, it is safe to say that chance and speculation are complementary companions. However, while there can be chance without speculation, there cannot be speculation without chance.

SOME TRUTHS, MYTHS, AND FALLACIES ABOUT STATISTICS

There are those who would give a loose definition of the social science discipline statistics as: "the art of selecting, organizing, and manipulating facts in such a manner as to infer from them whatever conclusions seem applicable at the moment."

Although it is essentially correct, the preceding definition of this complex, numerically based discipline is also elusive, incomplete, and even misleading. But such is the lot of statistics, a subject viewed by millions as a rather dreary business confined to the study of rather useless, complex, and often confusing data.

The end result of any statistical study is a set of conclusions about the organized data. These inferences are what a researcher, a politician, a businessman, an athlete, a union leader, and a psychologist (to name but a few users of statistics) apply to the future to help predict things to come individually and collectively. The inferences, the final steps in the chain of statistical reasoning and application, are totally dependent on the adequacy of the sample data that must be collected, presented, and analyzed. One faulty step and the conclusions become misleading and/or fallacious, if that is the researcher's real aim.

The science or art of statistical analysis can easily be compared to the construction of a house. Statistical research requires a design. What do you want to study . . . and why? The sample data corresponds to the builder's selection of raw materials. If the quality of the data is not specified beforehand, the research will crumble like the house built with substandard concrete in its foundation. So, like the builder or architect, the statistician must have a blueprint from which to work, for the quality and accuracy of the conclusions depends on the exactness of this formulation. Like a well-constructed house, statistical conclusions must stand up to scrutiniy and certification from those living with the data. As with a house, statistical tables, charts, and graphs are the results of experience and training. And as with a contractor, a statistician must stand by his efforts.

Before we discuss specific areas of interest where statistics play an important role, a list of possible applications for statistical research might prove enlightening. The list has been prepared randomly and it covers only those applications that came to mind over coffee on a rainy morning. The prescribed data is grouped with one possible application.

a) Car accidents—to determine safety requirements.
b) Burglaries—to calculate home insurance rates.
c) Reading level—to plan educational programs.
d) SAT scores—to correlate scores and a student's later grade point average in college.
e) Family incomes—to estimate a supermarket's sales.

f) Price fluctuations—to compute economic indicators.

g) Employment figures—to estimate the expected tax revenue.

h) Grain production—to estimate agricultural policy.

i) Life expectancy—to compute a pension system.

j) TV habits—to formulate programming.

k) Advertising evaluation—to plan an effective program.

l) Defective parts—to organize quality control.

m) Number of children—to determine number of schools needed.

n) Bus passenger traffic—to estimate future demand.

o) Inherited traits—to aid psychologists and sociologists.

p) Batting averages—to argue for salary demands.

q) Passing accuracy—to formulate a game plan.

r) The weight of women—to determine dress sizes.

s) Record sales—to estimate future production.

t) Accounting records—to formulate future business decisions.

u) The number of heart attacks—to determine future health policies.

v) Seasonal sales—to estimate future employment needs.

w) Weather patterns—to plan vacation dates.

x) Bathroom accidents—to reduce risks in the most dangerous room in the house.

y) Number of single women—to determine where a bachelor should live.

z) Banking records—to locate embezzlers before they leave for Brazil.

In the following sections, we will discuss specific areas in greater detail to investigate the uses, misuses, and abuses of statistical data.

Advertising

Without statistical surveys to verify its estimates and practices, the advertising industry would be like a man bereft of the senses of sight and hearing. Statistics provide the feedback that businesses require in order to plan a sales campaign.

Put simply, advertising's goal is to effect maximum results on a maximum number of people with a minimum of cost. When it works, advertising increases the net sales while it decreases the net expenses. The examples of companies that have succeeded without an effective advertising program can be counted on one, maybe two fingers, because, by its very nature, advertising brings a company or a product to the people.

In nearly every case, and whether the advertising medium employed is television, radio, newspapers, magazines, billboard, direct mail, T-shirts, or some exotic method like blimps or skywriting, the advertiser has to think in terms of the statistical group or the group of prospective customers to be reached. An estimate or measure of the value of the advertising choices and their potentials must be made prior to the start of the ad campaign.

The key is *demographics*, the science of the vital and social statistics—births, deaths, marriages, diseases, etc.—of a population. No advertising medium can survive unless it compiles demographics about its audience, for in cyclical fashion, the advertising media is dependent on advertising its wares.

The group of people exposed to any single advertising medium is generously comprised of smaller groups, each with a varying degree of value as prospective customers. Figure 9-B is an example of the composition of a group reached by a national woman's magazine. The composing of tables and charts enables an advertiser to pinpoint his market or target group.

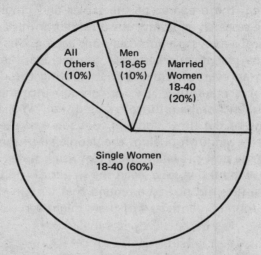

Figure 9-B. A pie chart for a women's magazines' readership.

The statistical problem for the prospective advertising medium and hence, the buyers of its advertising space, is to uncover the methods to dissect the entire population into its component groups, and then to find a numerical means of measuring the value of each in relation to the advertiser's goals.

If one is selling a product or service with mass across-the-board (demographic) appeal, the problem can be solved through a mass medium like television. On the other hand, if the product or service is aimed at one particular demographic group (i.e., men 18-35), the statistical measurement's accuracy is of great importance. In the former example, a shotgun is the appropriate weapon to hit the target. For the latter, one would require a bow and arrow and a steady hand, for a near-miss can prove as financially lethal as trying to get by without any advertising or promotion.

A secondary use of statistical analysis in advertising is for testing the effectiveness of one form of advertising against another. By formulating a null hypothesis based on the sample data received, an accurate estimate is definitely possible. The sample data can be obtained by asking customers to clip a coupon, call a specific telephone number, or write to a specific post office box.

The misuses and abuses of statistics in the field of advertising generally fall into 2 categories:

1) The statistical surveys are not prepared by professional statisticians.

Often, it is the marketing director or his assistant, without expertise, but with a modicum of layman's knowledge, who finds his facts and figures pliant and readily adaptable to his already biased viewpoint. Handling the numerical data with a cavalier ease, he tosses around a few measures of location and the sought-after conclusions will fairly leap off the table or graph at the prospective advertiser. Without bothering to learn about averages, dispersions, estimation, or hypothesis testing, the preparer of the tables and graphs is sure to oversimplify the research and draw unwarranted or misleading conclusions.

2) The statistical surveys are purposely misleading. The most misused and often abused statistical measure is the average, the arithmetic mean. By quoting an average without first quoting the data computed, an advertiser or a marketing director can convince the public of almost anything. When Abraham Lincoln stated that, "You can fool all of the people some of the time; you can fool some of the people all of the time; but you cannot fool all of the people all of the time," he naively formulated the now-prevalent theory of using statistical conclusions to mislead. How could anyone, even the writer of this book, argue that 4 out of 5 dentists did not recommend that certain sugarless gum? I'm sure if you found 5 dentists, 4 of them might agree on almost any conclusion.

Business

Are your earnings keeping pace with the annual cost-of-living index? Or, are you one of the 8+% of the work force that is presently unemployed?

Who are Dow Jones, Standard and Poors, and Dun and Bradstreet, and what do those statistical tables they produce mean to us?

Calvin Coolidge, our thirtieth president, was right when he stated: "The business of America is business." What Mr. Coolidge did not know was that the business of business has become a jumble of leading and misleading economic indicators, indices, averages, and statistical bookkeeping.

Today's businesses, from the international conglomerates to the neighborhood stores, provide a fertile, changing arena for statistical and probability theories. In business today, the large numbers of numerical facts, in diverse, complex patterns, require the employment of statistical analysis to just keep abreast and out of bankruptcy court. Statistics plays an important, integrated role in every business, profit or nonprofit, that is capable of handling large quantities of numbers. The easy access to small, personal computers and/or inexpensive data processing can put a small business on an equal footing with an industry giant if a business avails itself of the science of statistics.

Most of the facts a business handles are numerical. Everything that can be expressed in terms of dollars is numerical; for example, sales, purchases, expenses, accruals, dividends, and inventory. Everything that can be counted is likewise numerical and can be statistically measured; for example, numbers of sales, customers, employers, products manufactured, and on-the-job accidents.

In order for a business to understand what is happening in-house, throughout the industry, city-wide, state-wide, nationally, and internationally, it is essential that records are kept of many things that never appear on a company's financial statements. This expensive numerical data of every kind is fodder for the statistical research that forms the basis for the 6 leading economic indicators, the measurements compiled by the federal government to predict the direction in which the economy is moving. This data reflects trends in employment, productivity, interest rates (short-term and long-term), investments in new plant equipment, the sizes of retail inventories, and gross national product.

Manufacturers, the businesses that initiate the marketing chain that ends with wholesale or retail sales, were the first to understand and utilize the clear-cut advantages to be gained by applying modern statistical methods practiced by competent, scientifically trained researchers. A manufacturer's survival depends directly on scientific and engineering research, the measuring of supply and demand and market research opinion polls. Given a problem in the sphere of mathematical techniques, there is no other practical way to tackle it without statistics.

In the 1970s, the major United States auto companies had compiled enough statistical tables and analyses to have forecast the trends that are spelling economic disaster in the 1980s.

NOTE: In business, there are no secrets. Everything the Japanese or Germans have accomplished at our economic expense was and is well-documented statistically. Unlike theoretical mathematics, statistical theory is useless when practiced in a vacuum.

It is a pity that some businessmen still view the utilization of statistical research techniques as tantamount to abandoning seat-of-the-pants managerial judgment and good sense in favor of a set of rule-of-thumb procedures founded in statistics. The blind worship of traditional principles and practices is a serious obstacle to battling the competition in our high-tech world.

There should be only one overriding maxim: "Business research has to be 100% practical research which has to be justified in purely economic terms. It must be guided only by practical considerations."

Government and Politics

Nowhere is the utilization of statistical research more evident than in the complementary worlds of government and politics. Every politician, the "in" and the "wants-to-be-in," is spending thousands, sometimes millions of dollars (another statistical measurement worth studying) on radio and television commercials to quote statistics on: overspending, taxes, growth (or lack of), business starts, bankruptcies, research and development and those old statistical bugaboos, inflation (and the result of inflation), recession and crime.

How is it always possible for both sides to confidently quote statistics about the state of the government that totally contradict each other? The answer lies in the basis of the statistics, the data being analyzed and compared.

President Reagan stated that unemployment was up only 2.7% in the first two years of his Administration. His Democratic critics claimed that unemployment was up 35% since Reagan entered the White House. The contradictory statements are misleading, yet both offer conclusions that are correct. The confusion occurs because the percentages quoted were calculated in different ways. When the President took over in 1980, the unemployment rate was 7.7%. In the fall of 1982, the unemployment figure was 10.4%. If you subtract 7.7% from 10.4%, the difference is 2.7%. Thus, the President's statement is correct. However, if you use 7.7% as a base figure, the ratio of the 2.7% increase is 2.7%/7.7% or 35.1%. Thus, the President's critics have concluded correctly.

If you are still confused, you are not alone, because to judge the effectiveness of either statement, we must first know how the percentage of unemployment is determined and we must also know the definition of what is an unemployed person.

The calculation of unemployment is almost the same as the computation of a baseball player's batting average. In its simplest terms it is:

$$\frac{\text{the number of people without jobs}}{\text{the total number of people in the work force}}$$

At the end of 1975, the Labor Department stated that there were 7,717,257 unemployed out of a labor force of 92,979,000 Americans over the age of 16 who might be expected to work. The official computation was 8.3%. Unlike the computation of a batting average (total hits/total *official* at-bats), the calculation of unemployment figures is made vague in that the two basic terms are a bit ambiguous. Of the two, the concept of the labor force is only slightly more definite. Students are omitted during the school year, but they are included as "seasonal" workers during the summer. Housewives who work or have worked in the last 2 years are included, but other housewives who might be seeking employment are not, until they actively search for employment.

The definition of an unemployed person is more difficult. It would be simple just to include anyone looking for a job. However, this would also include anyone seeking to change jobs. Another definition might include only those unemployed people who have been out of work for a specified period of time, say 3 months. The first solution might have given us an unemployment rate of 11-12% in 1975. The second would have substantially lowered the rate from 8.3% to about 3-4%. The variables of unemployment are discrete, but very elusive. Is a part-time worker employed? What if he is seeking a full-time job? Is a student working part-time employed? What about a woman who is a student and a housewife, and works part-time?

What about the physically and mentally handicapped? Some are able to work. Some can only work part-time.

As you can see, the list could go on ad nauseum. However, the rate of unemployment is an important indicator. Many noted economists believe that an unemployment rate of over 9% indicates a period of recession. In 1975, if the almost one million students on work-study programs and citizens on government-created CETA jobs were not counted, the rate of unemployment would have approached 9.5%. It is the elusiveness of these indices that evokes the misstatements and half-truths found in political campaigns.

The opinion poll is another statistical technique that can foster fallacious or ambiguous statistical conclusions. Opinion polls are statistical predictions (inferences) based on a survey of a small sample. Such predictions necessarily involve error, and the smaller the sample, the larger the error. When the results of opinion polls, especially those prepared by companies hired by candidates, are offered as statistical evidence, the margin of error is usually omitted. The biased sample is another potential source of error in all surveys and especially in political opinion polls. To be as accurate as possible, the survey should be done in person. In surveys that are mailed, only a small percentage of those mailed are usually returned. In phone surveys, people tend to answer quickly to get off the phone. The bias occurs because there are large groups of people who never respond or who respond in haste. A third source of statistical fallacies is the time element. The rapidly changing conditions reported instantaneously on television and radio can make a survey outdated just days (even hours) after the statistics have been collected. Unlike economics statistical surveys that apply data which is quantitatively based, the polls used for political sampling generally employ qualitative variables. They generally divide into 2 categories: issues and personal reaction to the candidates.

The variables employed in 2 polling samples might include:

Issues	Personal Reactions
Crime (Gun Control)	Trustworthiness
Social Security	Experience
Taxes	Intelligence
Inflation	Party Loyalty
Unemployment	Charisma
Nuclear Weapons	Gender
Foreign Policy	
Education	

Many of the difficulties have been ironed out by the polls conducted by the mass media. The interviews are conducted in person by trained interviewers who ask precise questions of the unbiased, sufficiently large sample. Using computers to subject the resulting data to thorough statistical scrutiny, the data and the resulting inferences can be broadcast before the situation has time to change.

Because of the technology and the statistical expertise available today, it is highly unlikely that another polling debacle such as the prediction in 1948 that Thomas Dewey would soundly defeat Harry Truman could occur in 1982.

Yet, because it has become a common practice to attempt to find the answers to puzzling social, economic, and political questions by appealing for public opinion, abuses are possible. Magazines like *Cosmopolitan, Psychology Today, Playboy, Readers' Digest* and *Good Housekeeping* run reader mail-in surveys. The problems with mail-in as well as in-person surveys are obvious. Even if a person is puzzled by the question and the multiple-choice answers, he will answer and be listed with the minority or majority groups. The idea of government and political surveys seems to be that even if no interviewee knows the correct answer, if you ask enough individuals you will get a majority opinion to quote in all the mass media. It is human nature to give any answer rather than say "I don't know!"

Sports

Unlike the majority of fields that apply statistical methods practically, and a minority of fields that apply the procedures inordinately, the four major American team sports take the application of statistics to excess. One only had to watch the 1982 World Series between St. Louis and Milwaukee to witness the glut of statistics firsthand. (If I really cared to know how well Robin Yount hits against blonde, left-handed pitchers who were born in a state beginning with the letter "A," I would calculate it myself.)

Except for a few calculations, the statistics that fill the sports pages daily are computed by taking the number of successful attempts and dividing by the number of "official" attempts. The "official" comes into play when certain attempts in sports are not calculated into the total. For example, when computing a baseball player's batting average, the number of walks, sacrifices, hit-by-pitched-balls, catcher's interference, etc., are deduced from the total number of at-bats. In football, when calculating the efficiency of a passer or running back, the attempts that are nullified by a penalty are not computed. The same is true for hockey and basketball.

Two examples of sports statistics that are not computed in the aforementioned manner are the earned run average (ERA) for pitchers in baseball and the efficiency rating (plus and minus) for hockey players. To calculate a pitcher's ERA, the number of earned runs (without errors to assist the other team) the pitcher has allowed is divided by 9 to determine the ERA based on the standard nine-inning game. Thus, an ERA of 3.45 means the pitcher allows on the average of nearly 3-1/2 runs per every nine innings pitched. In hockey, a player's efficiency is measured by computing the number of goals scored by his team and by the opposing team while that player is on the ice (in even-strength situations). The player gets a plus each time he is on the ice when his team scores and a minus every time he is on the ice when the opposition scores. A player with more plusses than minuses is considered an asset, somewhere balanced offensively and defensively.

In all team and individual sports, statistics add zest and information. Properly used and understood, an individual or a team's statistics can prove an important instrument of strategy for the coach and player, as well as adding to a fan's enjoyment of the sport. When analyzed and applied properly, statistics can be converted in points (runs, goals, etc.) and points into victories for the individual, the team, the fan, and the sports bettor.

Although statistics can be formulated and computed for nearly every aspect of every sport, the statistics can *never* be the whole story of a single athlete's or a team's performance, and it would be a mistake for the opposition or a bettor to base strategy on the statistics alone. Statistics can indicate past history and specific tendencies in specific situations, but sports results, unlike results in other statistical arenas, are dependent on the interaction of human beings. Figures don't lie, but people can either soar to unexpected heights or fail in crucial situations. Without the unexpected and incalcuable human spirit, the games could just as easily be played like a real-life video game. Statistics can tell us what happened, but they cannot tell us why it happened.

Professional baseball was once called "a small island of physical activity played in a great ocean of facts and figures." I think the definition applies to all major team sports. Table 9-D offers a sample of statistics from the 4 major team sports in the United States. In the table, we will attempt to illustrate the most valuable and least valuable of the individual statistics.

Sport	Evaluation	Appropriate Statistics
I. Baseball	(1) Hitting	(1) Runs Scored (Speed)
		(2) Hits Per Game (Consistency)
		(3) RBI's/At Bats (Power)
		(4) Batting Average
	(2) Pitching	(1) ERA (Consistency, but the number of hits given is a better indicator)
		(2) Strikeouts/Inning Walk/Inning (Control)
		(3) Complete Games (for starting pitchers; shows durability)
		(4) Saves (for relief pitchers)
II. Football	(A) Offense	(A) Total Yards/Game
	(1) Rushing	(1) Yards/Game (Durability)
		(2) Yards/Carry (Consistency)
		(3) Fumbles (Tendencies)
	(2) Passing	(1) Yards/Game (Consistency)
		(2) Pass Completion Ratio
		(3) Total Yardage (Misleading statistic if team is way behind)
		(4) Interceptions/Attempts
	(B) Defense	(B) Yards/Per Game and Points/Game
	(1) Rushing	(1) Yards/Game (Consistency)
		(2) Yards/Carry (Tendencies)
		(3) Fumbles Recovered/Carry

Sport	Evaluation	Appropriate Statistics
	(2) Passing	(1) Yards/Game (Consistency) (2) Yards/Completion (Tendencies) (3) Interceptions/Attempts
	(C) Other	(C) Special Teams, Penalties, Intangibles (1) Yards/Kick (Punting) (2) Yards/Kick (Kickoff and Punt Returns) (3) Third Down Completions (Consistency) (4) Yards Penalized (Tendencies) (5) Time of Possession (Dominance)
III. Basketball	(1) Scoring	(1) Field Goal Percentage (Accuracy) (2) Free Throw Percentage (Pressure) (3) Assists/Game (Tendencies) (4) Points/Game (Consistency)
	(2) Rebounding	(1) Rebounds/Game (for starters) (2) Rebounds/Minutes (for reserves)
	(3) Defense	(1) Blocked Shots/Game (2) Steals/Game
IV. Hockey	(1) Scoring	(1) Shots on Goal (Consistency) (2) Goals (including short-handed goals) (3) Assists
	(2) Goaltending	(1) Goals Against Average (Consistency)
	(3) Defense and Others	(1) Plus and Minus

In individual sports, where the player is solely responsible for his (her) performance and hence, earnings, statistics are becoming more prevalent. Tennis players and coaches seeking an edge keenly check 2 statistics: Unforced errors (backhand, forehand, overhand, and volley) and the percentage of first serves in. Golfers who play a golf course's treacheries rather than an opponent have found statistics useful in evaluating their own strategies. The most useful seem to be: the number of putts per round and the number of greens reached in regulation per round.

Other sports, team and individual, have their own unique statistics. However, the sports and their statistical measurements are too numerous for inclusion here. Suffice it to say, every sport, big time or not, amateur or professional, has its statistics. Without them the only way to claim the distinction of superiority would be to beat the competition. What a silly notion. What would the World Series be without all the statistics? Who doesn't know and care that Reggie Jackson owns more expensive cars than any other left-handed hitting right fielder since 1923?

APPENDIX

FREQUENTLY USED STATISTICAL TABLES

Table 1
Binomial Distributions

Entries in the table are values of $\left(\dfrac{n}{x}\right) p^x (1-p)^{n-x}$ for the indicated values of n, x, and p. When $p > 0.5$, the value of $\left(\dfrac{n}{x}\right) p^x (1-p)^{n-x}$ for a given n, x, and p is obtained by finding the tabular entry for the given n, with $n - x$ in place of the given x, and $1 - p$ in place of the given p.

						p					
n	x	.05	.10	.15	.20	.25	.30	.35	.40	.45	.50
1	0	.9500	.9000	.8500	.8000	.7500	.7000	.6500	.6000	.5500	.5000
	1	.0500	.1000	.1500	.2000	.2500	.3000	.3500	.4000	.4500	.5000
2	0	.9025	.8100	.7225	.6400	.5625	.4900	.4225	.3600	.3025	.2500
	1	.0950	.1800	.2550	.3200	.3750	.4200	.4550	.4800	.4950	.5000
	2	.0025	.0100	.0225	.0400	.0625	.0900	.1225	.1600	.2025	.2500
3	0	.8574	.7290	.6141	.5120	.4219	.3430	.2746	.2160	.1664	.1250
	1	.1354	.2430	.3251	.3840	.4219	.4410	.4436	.4320	.4084	.3750
	2	.0071	.0270	.0574	.0960	.1406	.1890	.2389	.2880	.3341	.3750
	3	.0001	.0010	.0034	.0080	.0156	.0270	.0429	.0640	.0911	.1250
4	0	.8145	.6561	.5220	.4096	.3164	.2401	.1785	.1296	.0915	.0625
	1	.1715	.2916	.3685	.4096	.4219	.4116	.3845	.3456	.2995	.2500
	2	.0135	.0486	.0975	.1536	.2109	.2646	.3105	.3456	.3675	.3750
	3	.0005	.0036	.0115	.0256	.0469	.0756	.1115	.1536	.2005	.2500
	4	.0000	.0001	.0005	.0016	.0039	.0081	.0150	.0256	.0410	.0625
5	0	.7738	.5905	.4437	.3277	.2373	.1681	.1160	.0778	.0503	.0312
	1	.2036	.3280	.3915	.4096	.3955	.3602	.3124	.2592	.2059	.1562
	2	.0214	.0729	.1382	.2048	.2637	.3087	.3364	.3456	.3369	.3125
	3	.0011	.0081	.0244	.0512	.0879	.1323	.1811	.2304	.2757	.3125
	4	.0000	.0004	.0022	.0064	.0146	.0284	.0488	.0768	.1128	.1562
	5	.0000	.0000	.0001	.0003	.0010	.0024	.0053	.0102	.0185	.0312
6	0	.7351	.5314	.3771	.2621	.1780	.1176	.0754	.0467	.0277	.0156
	1	.2321	.3543	.3993	.3932	.3560	.3025	.2437	.1866	.1359	.0938
	2	.0305	.0984	.1762	.2458	.2966	.3241	.3280	.3110	.2780	.2344
	3	.0021	.0146	.0415	.0819	.1318	.1852	.2355	.2765	.3032	.3125
	4	.0001	.0012	.0055	.0154	.0330	.0595	.0951	.1382	.1861	.2344

Table 1 (continued)

n	x	.05	.10	.15	.20	.25	.30	.35	.40	.45	.50
6	5	.0000	.0001	.0004	.0015	.0044	.0102	.0205	.0369	.0609	.0938
	6	.0000	.0000	.0000	.0001	.0002	.0007	.0018	.0041	.0083	.0156
7	0	.6983	.4783	.3206	.2097	.1335	.0824	.0490	.0280	.0152	.0078
	1	.2573	.3720	.3960	.3670	.3115	.2471	.1848	.1306	.0872	.0547
	2	.0406	.1240	.2097	.2753	.3115	.3177	.2985	.2613	.2140	.1641
	3	.0036	.0230	.0617	.1147	.1730	.2269	.2679	.2903	.2918	.2734
	4	.0002	.0026	.0109	.0287	.0577	.0972	.1442	.1935	.2388	.2734
	5	.0000	.0002	.0012	.0043	.0115	.0250	.0466	.0774	.1172	.1641
	6	.0000	.0000	.0001	.0004	.0013	.0036	.0084	.0172	.0320	.0547
	7	.0000	.0000	.0000	.0000	.0001	.0002	.0006	.0016	.0037	.0078
8	0	.6634	.4305	.2725	.1678	.1001	.0576	.0319	.0168	.0084	.0039
	1	.2793	.3826	.3847	.3355	.2670	.1977	.1373	.0896	.0548	.0312
	2	.0515	.1488	.2376	.2936	.3115	.2965	.2587	.2090	.1569	.1094
	3	.0054	.0331	.0839	.1468	.2076	.2541	.2786	.2787	.2568	.2188
	4	.0004	.0046	.0185	.0459	.0865	.1361	.1875	.2322	.2627	.2734
	5	.0000	.0004	.0026	.0092	.0231	.0467	.0808	.1239	.1719	.2188
	6	.0000	.0000	.0002	.0011	.0038	.0100	.0217	.0413	.0703	.1094
	7	.0000	.0000	.0000	.0001	.0004	.0012	.0033	.0079	.0164	.0312
	8	.0000	.0000	.0000	.0000	.0000	.0001	.0002	.0007	.0017	.0039
9	0	.6302	.3874	.2316	.1342	.0751	.0404	.0207	.0101	.0046	.0020
	1	.2985	.3874	.3679	.3020	.2253	.1556	.1004	.0605	.0339	.0176
	2	.0629	.1722	.2597	.3020	.3003	.2668	.2162	.1612	.1110	.0703
	3	.0077	.0446	.1069	.1762	.2336	.2668	.2716	.2508	.2119	.1641
	4	.0006	.0074	.0283	.0661	.1168	.1715	.2194	.2508	.2600	.2461
	5	.0000	.0008	.0050	.0165	.0389	.0735	.1181	.1672	.2128	.2461
	6	.0000	.0001	.0006	.0028	.0087	.0210	.0424	.0743	.1160	.1641
	7	.0000	.0000	.0000	.0003	.0012	.0039	.0098	.0212	.0407	.0703
	8	.0000	.0000	.0000	.0000	.0001	.0004	.0013	.0035	.0083	.0176
	9	.0000	.0000	.0000	.0000	.0000	.0000	.0001	.0003	.0008	.0020
10	0	.5987	.3487	.1969	.1074	.0563	.0282	.0135	.0060	.0025	.0010
	1	.3151	.3874	.3474	.2684	.1877	.1211	.0725	.0403	.0207	.0098
	2	.0746	.1937	.2759	.3020	.2816	.2335	.1757	.1209	.0763	.0439
	3	.0105	.0574	.1298	.2013	.2503	.2668	.2522	.2150	.1665	.1172
	4	.0010	.0112	.0401	.0881	.1460	.2001	.2377	.2508	.2384	.2051
	5	.0001	.0015	.0085	.0264	.0584	.1029	.1536	.2007	.2340	.2461
	6	.0000	.0001	.0012	.0055	.0162	.0368	.0689	.1115	.1596	.2051
	7	.0000	.0000	.0001	.0008	.0031	.0090	.0212	.0425	.0746	.1172
	8	.0000	.0000	.0000	.0001	.0004	.0014	.0043	.0106	.0229	.0439
	9	.0000	.0000	.0000	.0000	.0000	.0001	.0005	.0016	.0042	.0098
	10	.0000	.0000	.0000	.0000	.0000	.0000	.0000	.0001	.0003	.0010

Table 1 (continued)

n	*x*	.05	.10	.15	.20	.25	.30	.35	.40	.45	.50
						p					
11	0	.5688	.3138	.1673	.0859	.0422	.0198	.0088	.0036	.0014	.0005
	1	.3293	.3835	.3248	.2362	.1549	.0932	.0518	.0266	.0125	.0054
	2	.0867	.2131	.2866	.2953	.2581	.1998	.1395	.0887	.0513	.0269
	3	.0137	.0710	.1517	.2215	.2581	.2568	.2254	.1774	.1259	.0806
	4	.0014	.0158	.0536	.1107	.1721	.2201	.2428	.2365	.2060	.1611
	5	.0001	.0025	.0132	.0388	.0803	.1321	.1830	.2207	.2360	.2256
	6	.0000	.0003	.0023	.0097	.0268	.0566	.0985	.1471	.1931	.2256
	7	.0000	.0000	.0003	.0017	.0064	.0173	.0379	.0701	.1128	.1611
	8	.0000	.0000	.0000	.0002	.0011	.0037	.0102	.0234	.0462	.0806
	9	.0000	.0000	.0000	.0000	.0001	.0005	.0018	.0052	.0126	.0269
	10	.0000	.0000	.0000	.0000	.0000	.0000	.0002	.0007	.0021	.0054
	11	.0000	.0000	.0000	.0000	.0000	.0000	.0000	.0000	.0002	.0005
12	0	.5404	.2824	.1422	.0687	.0317	.0138	.0057	.0022	.0008	.0002
	1	.3413	.3766	.3012	.2062	.1267	.0712	.0368	.0174	.0075	.0029
	2	.0988	.2301	.2924	.2835	.2323	.1678	.1088	.0639	.0339	.0161
	3	.0173	.0852	.1720	.2362	.2581	.2397	.1954	.1419	.0923	.0537
	4	.0021	.0213	.0683	.1329	.1936	.2311	.2367	.2128	.1700	.1208
	5	.0002	.0038	.0193	.0532	.1032	.1585	.2039	.2270	.2225	.1934
	6	.0000	.0005	.0040	.0155	.0401	.0792	.1281	.1766	.2124	.2256
	7	.0000	.0000	.0006	.0033	.0115	.0291	.0591	.1009	.1489	.1934
	8	.0000	.0000	.0001	.0005	.0024	.0078	.0199	.0420	.0762	.1208
	9	.0000	.0000	.0000	.0001	.0004	.0015	.0048	.0125	.0277	.0537
	10	.0000	.0000	.0000	.0000	.0000	.0002	.0008	.0025	.0068	.0161
	11	.0000	.0000	.0000	.0000	.0000	.0000	.0001	.0003	.0010	.0029
	12	.0000	.0000	.0000	.0000	.0000	.0000	.0000	.0000	.0001	.0002
13	0	.5133	.2542	.1209	.0550	.0238	.0097	.0037	.0013	.0004	.0001
	1	.3512	.3672	.2774	.1787	.1029	.0540	.0259	.0113	.0045	.0016
	2	.1109	.2448	.2937	.2680	.2059	.1388	.0836	.0453	.0220	.0095
	3	.0214	.0997	.1900	.2457	.2517	.2181	.1651	.1107	.0660	.0349
	4	.0028	.0277	.0838	.1535	.2097	.2337	.2222	.1845	.1350	.0873
	5	.0003	.0055	.0266	.0691	.1258	.1803	.2154	.2214	.1989	.1571
	6	.0000	.0008	.0063	.0230	.0559	.1030	.1546	.1968	.2169	.2095
	7	.0000	.0001	.0011	.0058	.0186	.0442	.0833	.1312	.1775	.2095
	8	.0000	.0000	.0001	.0011	.0047	.0142	.0336	.0656	.1089	.1571
	9	.0000	.0000	.0000	.0001	.0009	.0034	.0101	.0243	.0495	.0873
	10	.0000	.0000	.0000	.0000	.0001	.0006	.0022	.0065	.0162	.0349
	11	.0000	.0000	.0000	.0000	.0000	.0001	.0003	.0012	.0036	.0095
	12	.0000	.0000	.0000	.0000	.0000	.0000	.0000	.0001	.0005	.0016
	13	.0000	.0000	.0000	.0000	.0000	.0000	.0000	.0000	.0000	.0001

Table 1 (continued)

n	x	.05	.10	.15	.20	.25	.30	.35	.40	.45	.50
14	0	.4877	.2288	.1028	.0440	.0178	.0068	.0024	.0008	.0002	.0001
	1	.3593	.3559	.2539	.1539	.0832	.0407	.0181	.0073	.0027	.0009
	2	.1229	.2570	.2912	.2501	.1802	.1134	.0634	.0317	.0141	.0056
	3	.0259	.1142	.2056	.2501	.2402	.1943	.1366	.0845	.0462	.0222
	4	.0037	.0349	.0998	.1720	.2202	.2290	.2022	.1549	.1040	.0611
	5	.0004	.0078	.0352	.0860	.1468	.1963	.2178	.2066	.1701	.1222
	6	.0000	.0013	.0093	.0322	.0734	.1262	.1759	.2066	.2088	.1833
	7	.0000	.0002	.0019	.0092	.0280	.0618	.1082	.1574	.1952	.2095
	8	.0000	.0000	.0003	.0020	.0082	.0232	.0510	.0918	.1398	.1833
	9	.0000	.0000	.0000	.0003	.0018	.0066	.0183	.0408	.0762	.1222
	10	.0000	.0000	.0000	.0000	.0003	.0014	.0049	.0136	.0312	.0611
	11	.0000	.0000	.0000	.0000	.0000	.0002	.0010	.0033	.0093	.0222
	12	.0000	.0000	.0000	.0000	.0000	.0000	.0001	.0005	.0019	.0056
	13	.0000	.0000	.0000	.0000	.0000	.0000	.0000	.0001	.0002	.0009
	14	.0000	.0000	.0000	.0000	.0000	.0000	.0000	.0000	.0000	.0001
15	0	.4633	.2059	.0874	.0352	.0134	.0047	.0016	.0005	.0001	.0000
	1	.3658	.3432	.2312	.1319	.0668	.0305	.0126	.0047	.0016	.0005
	2	.1348	.2669	.2856	.2309	.1559	.0916	.0476	.0219	.0090	.0032
	3	.0307	.1285	.2184	.2501	.2252	.1700	.1110	.0634	.0318	.0139
	4	.0049	.0428	.1156	.1876	.2252	.2186	.1792	.1268	.0780	.0417
	5	.0006	.0105	.0449	.1032	.1651	.2061	.2123	.1859	.1404	.0916
	6	.0000	.0019	.0132	.0430	.0917	.1472	.1906	.2066	.1914	.1527
	7	.0000	.0003	.0030	.0138	.0393	.0811	.1319	.1771	.2013	.1964
	8	.0000	.0000	.0005	.0035	.0131	.0348	.0710	.1181	.1647	.1964
	9	.0000	.0000	.0001	.0007	.0034	.0116	.0298	.0612	.1048	.1527
	10	.0000	.0000	.0000	.0001	.0007	.0030	.0096	.0245	.0515	.0916
	11	.0000	.0000	.0000	.0000	.0001	.0006	.0024	.0074	.0191	.0417
	12	.0000	.0000	.0000	.0000	.0000	.0001	.0004	.0016	.0052	.0139
	13	.0000	.0000	.0000	.0000	.0000	.0000	.0001	.0003	.0010	.0032
	14	.0000	.0000	.0000	.0000	.0000	.0000	.0000	.0000	.0001	.0005
	15	.0000	.0000	.0000	.0000	.0000	.0000	.0000	.0000	.0000	.0000
16	0	.4401	.1853	.0743	.0281	.0100	.0033	.0010	.0003	.0001	.0000
	1	.3706	.3294	.2097	.1126	.0535	.0228	.0087	.0030	.0009	.0002
	2	.1463	.2745	.2775	.2111	.1336	.0732	.0353	.0150	.0056	.0018
	3	.0359	.1423	.2285	.2463	.2079	.1465	.0888	.0468	.0215	.0085
	4	.0061	.0514	.1311	.2001	.2252	.2040	.1553	.1014	.0572	.0278
	5	.0008	.0137	.0555	.1201	.1802	.2099	.2008	.1623	.1123	.0667
	6	.0001	.0028	.0180	.0550	.1101	.1649	.1982	.1983	.1684	.1222
	7	.0000	.0004	.0045	.0197	.0524	.1010	.1524	.1889	.1969	.1746
	8	.0000	.0001	.0009	.0055	.0197	.0487	.0923	.1417	.1812	.1964
	9	.0000	.0000	.0001	.0012	.0058	.0185	.0442	.0840	.1318	.1746

Table 1 (continued)

n	x	.05	.10	.15	.20	.25	.30	.35	.40	.45	.50
16	10	.0000	.0000	.0000	.0002	.0014	.0056	.0167	.0392	.0755	.1222
	11	.0000	.0000	.0000	.0000	.0002	.0013	.0049	.0142	.0337	.0667
	12	.0000	.0000	.0000	.0000	.0000	.0002	.0011	.0040	.0115	.0278
	13	.0000	.0000	.0000	.0000	.0000	.0000	.0002	.0008	.0029	.0085
	14	.0000	.0000	.0000	.0000	.0000	.0000	.0000	.0001	.0005	.0018
	15	.0000	.0000	.0000	.0000	.0000	.0000	.0000	.0000	.0001	.0002
	16	.0000	.0000	.0000	.0000	.0000	.0000	.0000	.0000	.0000	.0000
17	0	.4181	.1668	.0631	.0225	.0075	.0023	.0007	.0002	.0000	.0000
	1	.3741	.3150	.1893	.0957	.0426	.0169	.0060	.0019	.0005	.0001
	2	.1575	.2800	.2673	.1914	.1136	.0581	.0260	.0102	.0035	.0010
	3	.0415	.1556	.2359	.2393	.1893	.1245	.0701	.0341	.0144	.0052
	4	.0076	.0605	.1457	.2093	.2209	.1868	.1320	.0796	.0411	.0182
	5	.0010	.0175	.0668	.1361	.1914	.2081	.1849	.1379	.0875	.0472
	6	.0001	.0039	.0236	.0680	.1276	.1784	.1991	.1839	.1432	.0944
	7	.0000	.0007	.0065	.0267	.0668	.1201	.1685	.1927	.1841	.1484
	8	.0000	.0001	.0014	.0084	.0279	.0644	.1134	.1606	.1883	.1855
	9	.0000	.0000	.0003	.0021	.0093	.0276	.0611	.1070	.1540	.1855
	10	.0000	.0000	.0000	.0004	.0025	.0095	.0263	.0571	.1008	.1484
	11	.0000	.0000	.0000	.0001	.0005	.0026	.0090	.0242	.0525	.0944
	12	.0000	.0000	.0000	.0000	.0001	.0006	.0024	.0081	.0215	.0472
	13	.0000	.0000	.0000	.0000	.0000	.0001	.0005	.0021	.0068	.0182
	14	.0000	.0000	.0000	.0000	.0000	.0000	.0001	.0004	.0016	.0052
	15	.0000	.0000	.0000	.0000	.0000	.0000	.0000	.0001	.0003	.0010
	16	.0000	.0000	.0000	.0000	.0000	.0000	.0000	.0000	.0000	.0001
	17	.0000	.0000	.0000	.0000	.0000	.0000	.0000	.0000	.0000	.0000
18	0	.3972	.1501	.0536	.0180	.0056	.0016	.0004	.0001	.0000	.0000
	1	.3763	.3002	.1704	.0811	.0338	.0126	.0042	.0012	.0003	.0001
	2	.1683	.2835	.2556	.1723	.0958	.0458	.0190	.0069	.0022	.0006
	3	.0473	.1680	.2406	.2297	.1704	.1046	.0547	.0246	.0095	.0031
	4	.0093	.0700	.1592	.2153	.2130	.1681	.1104	.0614	.0291	.0117
	5	.0014	.0218	.0787	.1507	.1988	.2017	.1664	.1146	.0666	.0327
	6	.0002	.0052	.0301	.0816	.1436	.1873	.1941	.1655	.1181	.0708
	7	.0000	.0010	.0091	.0350	.0820	.1376	.1792	.1892	.1657	.1214
	8	.0000	.0002	.0022	.0120	.0376	.0811	.1327	.1734	.1864	.1669
	9	.0000	.0000	.0004	.0033	.0139	.0386	.0794	.1284	.1694	.1855
	10	.0000	.0000	.0001	.0008	.0042	.0149	.0385	.0771	.1248	.1669
	11	.0000	.0000	.0000	.0001	.0010	.0046	.0151	.0374	.0742	.1214
	12	.0000	.0000	.0000	.0000	.0002	.0012	.0047	.0145	.0354	.0708
	13	.0000	.0000	.0000	.0000	.0000	.0002	.0012	.0045	.0134	.0327
	14	.0000	.0000	.0000	.0000	.0000	.0000	.0002	.0011	.0039	.0117

Table 1 (continued)

n	x	.05	.10	.15	.20	.25	.30	.35	.40	.45	.50
18	15	.0000	.0000	.0000	.0000	.0000	.0000	.0000	.0002	.0009	.0031
	16	.0000	.0000	.0000	.0000	.0000	.0000	.0000	.0000	.0001	.0006
	17	.0000	.0000	.0000	.0000	.0000	.0000	.0000	.0000	.0000	.0001
	18	.0000	.0000	.0000	.0000	.0000	.0000	.0000	.0000	.0000	.0000
19	0	.3774	.1351	.0456	.0144	.0042	.0011	.0003	.0001	.0000	.0000
	1	.3774	.2852	.1529	.0685	.0268	.0093	.0029	.0008	.0002	.0000
	2	.1787	.2852	.2428	.1540	.0803	.0358	.0138	.0046	.0013	.0003
	3	.0533	.1796	.2428	.2182	.1517	.0869	.0422	.0175	.0062	.0018
	4	.0112	.0798	.1714	.2182	.2023	.1491	.0909	.0467	.0203	.0074
	5	.0018	.0266	.0907	.1636	.2023	.1916	.1468	.0933	.0497	.0222
	6	.0002	.0069	.0374	.0955	.1574	.1916	.1844	.1451	.0949	.0518
	7	.0000	.0014	.0122	.0443	.0974	.1525	.1844	.1797	.1443	.0961
	8	.0000	.0002	.0032	.0166	.0487	.0981	.1489	.1797	.1771	.1442
	9	.0000	.0000	.0007	.0051	.0198	.0514	.0980	.1464	.1771	.1762
	10	.0000	.0000	.0001	.0013	.0066	.0220	.0528	.0976	.1449	.1762
	11	.0000	.0000	.0000	.0003	.0018	.0077	.0233	.0532	.0970	.1442
	12	.0000	.0000	.0000	.0000	.0004	.0022	.0083	.0237	.0529	.0961
	13	.0000	.0000	.0000	.0000	.0001	.0005	.0024	.0085	.0233	.0518
	14	.0000	.0000	.0000	.0000	.0000	.0001	.0006	.0024	.0082	.0222
	15	.0000	.0000	.0000	.0000	.0000	.0000	.0001	.0005	.0022	.0074
	16	.0000	.0000	.0000	.0000	.0000	.0000	.0000	.0001	.0005	.0018
	17	.0000	.0000	.0000	.0000	.0000	.0000	.0000	.0000	.0001	.0003
	18	.0000	.0000	.0000	.0000	.0000	.0000	.0000	.0000	.0000	.0000
	19	.0000	.0000	.0000	.0000	.0000	.0000	.0000	.0000	.0000	.0000
20	0	.3585	.1216	.0388	.0115	.0032	.0008	.0002	.0000	.0000	.0000
	1	.3774	.2702	.1368	.0576	.0211	.0068	.0020	.0005	.0001	.0000
	2	.1887	.2852	.2293	.1369	.0669	.0278	.0100	.0031	.0008	.0002
	3	.0596	.1901	.2428	.2054	.1339	.0716	.0323	.0123	.0040	.0011
	4	.0133	.0898	.1821	.2182	.1897	.1304	.0738	.0350	.0139	.0046
	5	.0022	.0319	.1028	.1746	.2023	.1789	.1272	.0746	.0365	.0148
	6	.0003	.0089	.0454	.1091	.1686	.1916	.1712	.1244	.0746	.0370
	7	.0000	.0020	.0160	.0545	.1124	.1643	.1844	.1659	.1221	.0739
	8	.0000	.0004	.0046	.0222	.0609	.1144	.1614	.1797	.1623	.1201
	9	.0000	.0001	.0011	.0074	.0271	.0654	.1158	.1597	.1771	.1602
	10	.0000	.0000	.0002	.0020	.0099	.0308	.0686	.1171	.1593	.1762
	11	.0000	.0000	.0000	.0005	.0030	.0120	.0336	.0710	.1185	.1602
	12	.0000	.0000	.0000	.0001	.0008	.0039	.0136	.0355	.0727	.1201
	13	.0000	.0000	.0000	.0000	.0002	.0010	.0045	.0146	.0366	.0739
	14	.0000	.0000	.0000	.0000	.0000	.0002	.0012	.0049	.0150	.0370

Table 1 (continued)

n	x					p					
		.05	.10	.15	.20	.25	.30	.35	.40	.45	.50
20	15	.0000	.0000	.0000	.0000	.0000	.0000	.0003	.0013	.0049	.0148
	16	.0000	.0000	.0000	.0000	.0000	.0000	.0000	.0003	.0013	.0046
	17	.0000	.0000	.0000	.0000	.0000	.0000	.0000	.0000	.0002	.0011
	18	.0000	.0000	.0000	.0000	.0000	.0000	.0000	.0000	.0000	.0002
	19	.0000	.0000	.0000	.0000	.0000	.0000	.0000	.0000	.0000	.0000
	20	.0000	.0000	.0000	.0000	.0000	.0000	.0000	.0000	.0000	.0000

Table 2
Normal Distribution

Example:
If $z = 1.96$, then
$P(0 \text{ to } z) = 0.4750$

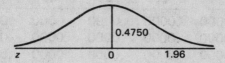

0.4750

z   0   1.96

Areas Under the Normal Curve

z	0.00	0.01	0.02	0.03	0.04	0.05	0.06	0.07	0.08	0.09
0.0	0.0000	0.0040	0.0080	0.0120	0.0160	0.0199	0.0239	0.0279	0.0319	0.0359
0.1	0.0398	0.0438	0.0478	0.0517	0.0557	0.0596	0.0636	0.0675	0.0714	0.0753
0.2	0.0793	0.0832	0.0871	0.0910	0.0948	0.0987	0.1026	0.1064	0.1103	0.1141
0.3	0.1179	0.1217	0.1255	0.1293	0.1331	0.1368	0.1406	0.1443	0.1480	0.1517
0.4	0.1554	0.1591	0.1628	0.1664	0.1700	0.1736	0.1772	0.1808	0.1844	0.1879
0.5	0.1915	0.1950	0.1985	0.2019	0.2054	0.2088	0.2123	0.2157	0.2190	0.2224
0.6	0.2257	0.2291	0.2324	0.2357	0.2389	0.2422	0.2454	0.2486	0.2517	0.2549
0.7	0.2580	0.2611	0.2642	0.2673	0.2704	0.2734	0.2764	0.2794	0.2823	0.2852
0.8	0.2881	0.2910	0.2939	0.2967	0.2995	0.3023	0.3051	0.3078	0.3106	0.3133
0.9	0.3159	0.3186	0.3212	0.3238	0.3264	0.3289	0.3315	0.3340	0.3365	0.3389
1.0	0.3413	0.3438	0.3461	0.3485	0.3508	0.3531	0.3554	0.3577	0.3599	0.3621
1.1	0.3643	0.3665	0.3686	0.3708	0.3729	0.3749	0.3770	0.3790	0.3810	0.3830
1.2	0.3849	0.3869	0.3888	0.3907	0.3925	0.3944	0.3962	0.3980	0.3997	0.4015
1.3	0.4032	0.4049	0.4066	0.4082	0.4099	0.4115	0.4131	0.4147	0.4162	0.4177
1.4	0.4192	0.4207	0.4222	0.4236	0.4251	0.4265	0.4279	0.4292	0.4306	0.4319
1.5	0.4332	0.4345	0.4357	0.4370	0.4382	0.4394	0.4406	0.4418	0.4429	0.4441
1.6	0.4452	0.4463	0.4474	0.4484	0.4495	0.4505	0.4515	0.4525	0.4535	0.4545
1.7	0.4554	0.4564	0.4573	0.4582	0.4591	0.4599	0.4608	0.4616	0.4625	0.4633
1.8	0.4641	0.4649	0.4656	0.4664	0.4671	0.4678	0.4686	0.4693	0.4699	0.4706
1.9	0.4713	0.4719	0.4726	0.4732	0.4738	0.4744	0.4750	0.4756	0.4761	0.4767
2.0	0.4772	0.4778	0.4783	0.4788	0.4793	0.4798	0.4803	0.4808	0.4812	0.4817
2.1	0.4821	0.4826	0.4830	0.4834	0.4838	0.4842	0.4846	0.4850	0.4854	0.4857
2.2	0.4861	0.4864	0.4868	0.4871	0.4875	0.4878	0.4881	0.4884	0.4887	0.4890
2.3	0.4893	0.4896	0.4898	0.4901	0.4904	0.4906	0.4909	0.4911	0.4913	0.4916
2.4	0.4918	0.4920	0.4922	0.4925	0.4927	0.4929	0.4931	0.4932	0.4934	0.4936
2.5	0.4938	0.4940	0.4941	0.4943	0.4945	0.4946	0.4948	0.4949	0.4951	0.4952
2.6	0.4953	0.4955	0.4956	0.4957	0.4959	0.4960	0.4961	0.4962	0.4963	0.4964
2.7	0.4965	0.4966	0.4967	0.4968	0.4969	0.4970	0.4971	0.4972	0.4973	0.4974
2.8	0.4974	0.4975	0.4976	0.4977	0.4977	0.4978	0.4979	0.4979	0.4980	0.4981
2.9	0.4981	0.4982	0.4982	0.4983	0.4984	0.4984	0.4985	0.4985	0.4986	0.4986
3.0	0.4987	0.4987	0.4987	0.4988	0.4988	0.4989	0.4989	0.4989	0.4990	0.4990

Table 3
The Student τ-Distribution

Level of Significance for 1-Tailed Test

	.10	.05	.025	.01	.005	.0005

Level of Significance for 2-Tailed Test

df	.20	.10	.05	.02	.01	.001
1	3.078	6.314	12.706	31.821	63.657	636.619
2	1.886	2.920	4.303	6.965	9.925	31.598
3	1.638	2.353	3.182	4.541	5.841	12.941
4	1.533	2.132	2.776	3.747	4.604	8.610
5	1.476	2.015	2.571	3.365	4.032	6.859
6	1.440	1.943	2.447	3.143	3.707	5.959
7	1.415	1.895	2.365	2.998	3.499	5.405
8	1.397	1.860	2.306	2.896	3.355	5.041
9	1.383	1.833	2.262	2.821	3.250	4.781
10	1.372	1.812	2.228	2.764	3.169	4.587
11	1.363	1.796	2.201	2.718	3.106	4.437
12	1.356	1.782	2.179	2.681	3.055	4.318
13	1.350	1.771	2.160	2.650	3.012	4.221
14	1.345	1.761	2.145	2.624	2.977	4.140
15	1.341	1.753	2.131	2.602	2.947	4.073
16	1.337	1.746	2.120	2.583	2.921	4.015
17	1.333	1.740	2.110	2.567	2.898	3.965
18	1.330	1.734	2.101	2.552	2.878	3.922
19	1.328	1.729	2.093	2.539	2.861	3.883
20	1.325	1.725	2.086	2.528	2.845	3.850
21	1.323	1.721	2.080	2.518	2.831	3.819
22	1.321	1.717	2.074	2.508	2.819	3.792
23	1.319	1.714	2.069	2.500	2.807	3.767
24	1.318	1.711	2.064	2.492	2.797	3.745
25	1.316	1.708	2.060	2.485	2.787	3.725
26	1.315	1.706	2.056	2.479	2.779	3.707
27	1.314	1.703	2.052	2.473	2.771	3.690
28	1.313	1.701	2.048	2.467	2.763	3.674
29	1.311	1.699	2.045	2.462	2.756	3.659
30	1.310	1.697	2.042	2.457	2.750	3.646
40	1.303	1.684	2.021	2.423	2.704	3.551
60	1.296	1.671	2.000	2.390	2.660	3.460
120	1.289	1.658	1.980	2.358	2.617	3.373
∞	1.282	1.645	1.960	2.326	2.576	3.291

Table 4
The Upper Percentage Points of the Chi-Square (χ^2) Distribution

df	.99	.98	.95	.90	.80	.70	.50	.30	.20	.10	.05	.02	.01	.001
1	$0.^3157$	$0.^3628$	.00393	.0158	.0642	.148	.455	1.074	1.642	2.706	3.841	5.412	6.635	10.827
2	.0201	.0404	.103	.211	.446	.713	1.386	2.408	3.219	4.605	5.991	7.824	9.210	13.815
3	.115	.185	.352	.584	1.005	1.424	2.366	3.665	4.642	6.251	7.815	9.837	11.345	16.266
4	.297	.429	.711	1.064	1.649	2.195	3.357	4.878	5.989	7.779	9.488	11.668	13.277	18.467
5	.554	.752	1.145	1.610	2.343	3.000	4.351	6.064	7.289	9.236	11.070	13.388	15.086	20.515
6	.872	1.134	1.635	2.204	3.070	3.828	5.348	7.231	8.558	10.645	12.592	15.033	16.812	22.457
7	1.239	1.564	2.167	2.833	3.822	4.671	6.346	8.383	9.803	12.017	14.067	16.622	18.475	24.322
8	1.646	2.032	2.733	3.490	4.594	5.527	7.344	9.524	11.030	13.362	15.507	18.168	20.090	26.125
9	2.088	2.532	3.325	4.168	5.380	6.393	8.343	10.656	12.242	14.684	16.919	19.679	21.666	27.877
10	2.558	3.059	3.940	4.865	6.179	7.267	9.342	11.781	13.442	15.987	18.307	21.161	23.209	29.588

11	3.053	3.609	4.575	5.578	6.989	8.148	10.341	12.899	14.631	17.275	19.675	22.618	24.725	31.264
12	3.571	4.178	5.226	6.304	7.807	9.034	11.340	14.011	15.812	18.549	21.026	24.054	26.217	32.909
13	4.107	4.765	5.892	7.042	8.634	9.926	12.340	15.119	16.985	19.812	22.362	25.472	27.688	34.528
14	4.660	5.368	6.571	7.790	9.467	10.821	13.339	16.222	18.151	21.064	23.685	26.873	29.141	36.123
15	5.229	5.985	7.261	8.547	10.307	11.721	14.339	17.322	19.311	22.307	24.996	28.259	30.578	37.697
16	5.812	6.614	7.962	9.312	11.152	12.624	15.338	18.418	20.465	23.542	26.296	29.633	32.000	39.252
17	6.408	7.255	8.672	10.085	12.002	13.531	16.338	19.511	21.615	24.769	27.587	30.995	33.409	40.790
18	7.015	7.906	9.390	10.865	12.857	14.440	17.338	20.601	22.760	25.989	28.869	32.346	34.805	42.312
19	7.633	8.567	10.117	11.651	13.716	15.352	18.338	21.689	23.900	27.204	30.144	33.687	36.191	43.820
20	8.260	9.237	10.851	12.443	14.578	16.266	19.337	22.775	25.038	28.412	31.410	35.020	37.566	45.315
21	8.897	9.915	11.591	13.240	15.445	17.182	20.337	23.858	26.171	29.615	32.671	36.343	38.932	46.797
22	9.542	10.600	12.338	14.041	16.314	18.101	21.337	24.939	27.301	30.813	33.924	37.659	40.289	48.268
23	10.196	11.293	13.091	14.848	17.187	19.021	22.337	26.018	28.429	32.007	35.172	38.968	41.638	49.728
24	10.856	11.992	13.848	15.659	18.062	19.943	23.337	27.096	29.553	33.196	36.415	40.270	42.980	51.179
25	11.524	12.697	14.611	16.473	18.940	20.867	24.337	28.172	30.675	34.382	37.652	41.566	44.314	52.620
26	12.198	13.409	15.379	17.292	19.820	21.792	25.336	29.246	31.795	35.563	38.885	42.856	45.642	54.052
27	12.879	14.125	16.151	18.114	20.703	22.719	26.336	30.319	32.912	36.741	40.113	44.140	46.963	55.476
28	13.565	14.847	16.928	18.939	21.588	23.647	27.336	31.391	34.027	37.916	41.337	45.419	43.278	56.893
29	14.256	15.574	17.708	19.768	22.475	24.577	28.336	32.461	35.139	39.087	42.557	46.693	49.588	58.302
30	14.953	16.306	18.493	20.599	23.364	25.508	29.336	33.530	36.250	40.256	43.773	47.962	50.892	59.703

Table 5
The Upper Percentage Points for the f-Distribution

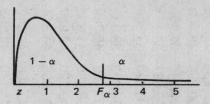

df for de-nomi-nator	α	\multicolumn{12}{c}{df for Numerator}
		1	2	3	4	5	6	7	8	9	10	11	12
1	.25	5.83	7.50	8.20	8.58	8.82	8.98	9.10	9.19	9.26	9.32	9.36	9.41
	.10	39.9	49.5	53.6	55.8	57.2	58.2	58.9	59.4	59.9	60.2	60.5	60.7
	.05	161	200	216	225	230	234	237	239	241	242	243	244
2	.25	2.57	3.00	3.15	3.23	3.28	3.31	3.34	3.35	3.37	3.38	3.39	3.39
	.10	8.53	9.00	9.16	9.24	9.29	9.33	9.35	9.37	9.38	9.39	9.40	9.41
	.05	18.5	19.0	19.2	19.2	19.3	19.3	19.4	19.4	19.4	19.4	19.4	19.4
	.01	98.5	99.0	99.2	99.2	99.3	99.3	99.4	99.4	99.4	99.4	99.4	99.4
3	.25	2.02	2.28	2.36	2.39	2.41	2.42	2.43	2.44	2.44	2.44	2.45	2.45
	.10	5.54	5.46	5.39	5.34	5.31	5.28	5.27	5.25	5.24	5.23	5.22	5.22
	.05	10.1	9.55	9.28	9.12	9.01	8.94	8.89	8.85	8.81	8.79	8.76	8.74
	.01	34.1	30.8	29.5	28.7	28.2	27.9	27.7	27.5	27.3	27.2	27.1	27.1
4	.25	1.81	2.00	2.05	2.06	2.07	2.08	2.08	2.08	2.08	2.08	2.08	2.08
	.10	4.54	4.32	4.19	4.11	4.05	4.01	3.98	3.95	3.94	3.92	3.91	3.90
	.05	7.71	6.94	6.59	6.39	6.26	6.16	6.09	6.04	6.00	5.96	5.94	5.91
	.01	21.2	18.0	16.7	16.0	15.5	15.2	15.0	14.8	14.7	14.5	14.4	14.4
5	.25	1.69	1.85	1.88	1.89	1.89	1.89	1.89	1.89	1.89	1.89	1.89	1.89
	.10	4.06	3.78	3.62	3.52	3.45	3.40	3.37	3.34	3.32	3.30	3.28	3.27
	.05	6.61	5.79	5.41	5.19	5.05	4.95	4.88	4.82	4.77	4.74	4.71	4.68
	.01	16.3	13.3	12.1	11.4	11.0	10.7	10.5	10.3	10.2	10.1	9.96	9.89
6	.25	1.62	1.76	1.78	1.79	1.79	1.78	1.78	1.78	1.77	1.77	1.77	1.77
	.10	3.78	3.46	3.29	3.18	3.11	3.05	3.01	2.98	2.96	2.94	2.92	2.90
	.05	5.99	5.14	4.76	4.53	4.39	4.28	4.21	4.15	4.10	4.06	4.03	4.00
	.01	13.7	10.9	9.78	9.15	8.75	8.47	8.26	8.10	7.98	7.87	7.79	7.72
7	.25	1.57	1.70	1.72	1.72	1.71	1.71	1.70	1.70	1.69	1.69	1.69	1.68
	.10	3.59	3.26	3.07	2.96	2.88	2.83	2.78	2.75	2.72	2.70	2.68	2.67
	.05	5.59	4.74	4.35	4.12	3.97	3.87	3.79	3.73	3.68	3.64	3.60	3.57
	.01	12.2	9.55	8.45	7.85	7.46	7.19	6.99	6.84	6.72	6.62	6.54	6.47
8	.25	1.54	1.66	1.67	1.66	1.66	1.65	1.64	1.64	1.63	1.63	1.63	1.62
	.10	3.46	3.11	2.92	2.81	2.73	2.67	2.62	2.59	2.56	2.54	2.52	2.50
	.05	5.32	4.46	4.07	3.84	3.69	3.58	3.50	3.44	3.39	3.35	3.31	3.28
	.01	11.3	8.65	7.59	7.01	6.63	6.37	6.18	6.03	5.91	5.81	5.73	5.67
9	.25	1.51	1.62	1.63	1.63	1.62	1.61	1.60	1.60	1.59	1.59	1.58	1.58
	.10	3.36	3.01	2.81	2.69	2.61	2.55	2.51	2.47	2.44	2.42	2.40	2.38
	.05	5.12	4.26	3.86	3.63	3.48	3.37	3.29	3.23	3.18	3.14	3.10	3.07
	.01	10.6	8.02	6.99	6.42	6.06	5.80	5.61	5.47	5.35	5.26	5.18	5.11

df for Numerator

15	20	24	30	40	50	60	100	120	200	500	∞
9.49	9.58	9.63	9.67	9.71	9.74	9.76	9.78	9.80	9.82	9.84	9.85
61.2	61.7	62.0	62.3	62.5	62.7	62.8	63.0	63.1	63.2	63.3	63.3
246	248	249	250	251	252	252	253	253	254	254	254
3.41	3.43	3.43	3.44	3.45	3.45	3.46	3.47	3.47	3.48	3.48	3.48
9.42	9.44	9.45	9.46	9.47	9.47	9.47	9.48	9.48	9.49	9.49	9.49
19.4	19.4	19.5	19.5	19.5	19.5	19.5	19.5	19.5	19.5	19.5	19.5
99.4	99.4	99.5	99.5	99.5	99.5	99.5	99.5	99.5	99.5	99.5	99.5
2.46	2.46	2.46	2.47	2.47	2.47	2.47	2.47	2.47	2.47	2.47	2.47
5.20	5.18	5.18	5.17	5.16	5.15	5.15	5.14	5.14	5.14	5.14	5.13
8.70	8.66	8.64	8.62	8.59	8.58	8.57	8.55	8.55	8.54	8.53	8.53
26.9	26.7	26.6	26.5	26.4	26.4	26.3	26.2	26.2	26.2	26.1	26.1
2.08	2.08	2.08	2.08	2.08	2.08	2.08	2.08	2.08	2.08	2.08	2.08
3.87	3.84	3.83	3.82	3.80	3.80	3.79	3.78	3.78	3.77	3.76	3.76
5.86	5.80	5.77	5.75	5.72	5.70	5.69	5.66	5.66	5.65	5.64	5.63
14.2	14.0	13.9	13.8	13.7	13.7	13.7	13.6	13.6	13.5	13.5	13.5
1.89	1.88	1.88	1.88	1.88	1.88	1.87	1.87	1.87	1.87	1.87	1.87
3.24	3.21	3.19	3.17	3.16	3.15	3.14	3.13	3.12	3.12	3.11	3.10
4.62	4.56	4.53	4.50	4.46	4.44	4.43	4.41	4.40	4.39	4.37	4.36
9.72	9.55	9.47	9.38	9.29	9.24	9.20	9.13	9.11	9.08	9.04	9.02
1.76	1.76	1.75	1.75	1.75	1.75	1.74	1.74	1.74	1.74	1.74	1.74
2.87	2.84	2.82	2.80	2.78	2.77	2.76	2.75	2.74	2.73	2.73	2.72
3.94	3.87	3.84	3.81	3.77	3.75	3.74	3.71	3.70	3.69	3.68	3.67
7.56	7.40	7.31	7.23	7.14	7.09	7.06	6.99	6.97	6.93	6.90	6.88
1.68	1.67	1.67	1.66	1.66	1.66	1.65	1.65	1.65	1.65	1.65	1.65
2.63	2.59	2.58	2.56	2.54	2.52	2.51	2.50	2.49	2.48	2.48	2.47
3.51	3.44	3.41	3.38	3.34	3.32	3.30	3.27	3.27	3.25	3.24	3.23
6.31	6.16	6.07	5.99	5.91	5.86	5.82	5.75	5.74	5.70	5.67	5.65
1.62	1.61	1.60	1.60	1.59	1.59	1.59	1.58	1.58	1.58	1.58	1.58
2.46	2.42	2.40	2.38	2.36	2.35	2.34	2.32	2.32	2.31	2.30	2.29
3.22	3.15	3.12	3.08	3.04	3.02	3.01	2.97	2.97	2.95	2.94	2.93
5.52	5.36	5.28	5.20	5.12	5.07	5.03	4.96	4.95	4.91	4.88	4.86
1.57	1.56	1.56	1.55	1.55	1.54	1.54	1.53	1.53	1.53	1.53	1.53
2.34	2.30	2.28	2.25	2.23	2.22	2.21	2.19	2.18	2.17	2.17	2.16
3.01	2.94	2.90	2.86	2.83	2.80	2.79	2.76	2.75	2.73	2.72	2.71
4.96	4.81	4.73	4.65	4.57	4.52	4.48	4.42	4.40	4.36	4.33	4.31

Table 5 (continued)

df for de-nomi-nator	α	\	df for Numerator										
		1	2	3	4	5	6	7	8	9	10	11	12
	.25	1.49	1.60	1.60	1.59	1.59	1.58	1.57	1.56	1.56	1.55	1.55	1.54
10	.10	3.29	2.92	2.73	2.61	2.52	2.46	2.41	2.38	2.35	2.32	2.30	2.28
	.05	4.96	4.10	3.71	3.48	3.33	3.22	3.14	3.07	3.02	2.98	2.94	2.91
	.01	10.0	7.56	6.55	5.99	5.64	5.39	5.20	5.06	4.94	4.85	4.77	4.71
	.25	1.47	1.58	1.58	1.57	1.56	1.55	1.54	1.53	1.53	1.52	1.52	1.51
11	.10	3.23	2.86	2.66	2.54	2.45	2.39	2.34	2.30	2.27	2.25	2.23	2.21
	.05	4.84	3.98	3.59	3.36	3.20	3.09	3.01	2.95	2.90	2.85	2.82	2.79
	.01	9.65	7.21	6.22	5.67	5.32	5.07	4.89	4.74	4.63	4.54	4.46	4.40
	.25	1.46	1.56	1.56	1.55	1.54	1.53	1.52	1.51	1.51	1.50	1.50	1.49
12	.10	3.18	2.81	2.61	2.48	2.39	2.33	2.28	2.24	2.21	2.19	2.17	2.15
	.05	4.75	3.89	3.49	3.26	3.11	3.00	2.91	2.85	2.80	2.75	2.72	2.69
	.01	9.33	6.93	5.95	5.41	5.06	4.82	4.64	4.50	4.39	4.30	4.22	4.16
	.25	1.45	1.55	1.55	1.53	1.52	1.51	1.50	1.49	1.49	1.48	1.47	1.47
13	.10	3.14	2.65	2.56	2.43	2.35	2.28	2.23	2.20	2.16	2.14	2.12	2.10
	.05	4.67	3.81	3.41	3.18	3.03	2.92	2.83	2.77	2.71	2.67	2.63	2.60
	.01	9.07	6.70	5.74	5.21	4.86	4.62	4.44	4.30	4.19	4.10	4.02	3.96
	.25	1.44	1.53	1.53	1.52	1.51	1.50	1.49	1.48	1.47	1.46	1.46	1.45
14	.10	3.10	2.73	2.52	2.39	2.31	2.24	2.19	2.15	2.12	2.10	2.08	2.05
	.05	4.60	3.74	3.34	3.11	2.96	2.85	2.76	2.70	2.65	2.60	2.57	2.53
	.01	8.86	6.51	5.56	5.04	4.69	4.46	4.28	4.14	4.03	3.94	3.86	3.80
	.25	1.43	1.52	1.52	1.51	1.49	1.48	1.47	1.46	1.46	1.45	1.44	1.44
15	.10	3.07	2.70	2.49	2.36	2.27	2.21	2.16	2.12	2.09	2.06	2.04	2.02
	.05	4.54	3.68	3.29	3.06	2.90	2.79	2.71	2.64	2.59	2.54	2.51	2.48
	.01	8.68	6.36	5.42	4.89	4.56	4.32	4.14	4.00	3.89	3.80	3.73	3.67
	.25	1.42	1.51	1.51	1.50	1.48	1.47	1.46	1.45	1.44	1.44	1.44	1.43
16	.10	3.05	2.67	2.46	2.33	2.24	2.18	2.13	2.09	2.06	2.03	2.01	1.99
	.05	4.49	3.63	3.24	3.01	2.85	2.74	2.66	2.59	2.54	2.49	2.46	2.42
	.01	8.53	6.23	5.29	4.77	4.44	4.20	4.03	3.89	3.78	3.69	3.62	3.55
	.25	1.42	1.51	1.50	1.49	1.47	1.46	1.45	1.44	1.43	1.43	1.42	1.41
17	.10	3.03	2.64	2.44	2.31	2.22	2.15	2.10	2.06	2.03	2.00	1.98	1.96
	.05	4.45	3.59	3.20	2.96	2.81	2.70	2.61	2.55	2.49	2.45	2.41	2.38
	.01	8.40	6.11	5.18	4.67	4.34	4.10	3.93	3.79	3.68	3.59	3.52	3.46
	.25	1.41	1.50	1.49	1.48	1.46	1.45	1.44	1.43	1.42	1.42	1.41	1.40
18	.10	3.01	2.62	2.42	2.29	2.20	2.13	2.08	2.04	2.00	1.98	1.96	1.93
	.05	4.41	3.55	3.16	2.93	2.77	2.66	2.58	2.51	2.46	2.41	2.37	2.34
	.01	8.29	6.01	5.09	4.58	4.25	4.01	3.84	3.71	3.60	3.51	3.43	3.37
	.25	1.41	1.49	1.49	1.47	1.46	1.44	1.43	1.42	1.41	1.41	1.40	1.40
19	.10	2.99	2.61	2.40	2.27	2.18	2.11	2.06	2.02	1.98	1.96	1.94	1.91
	.05	4.38	3.52	3.13	2.90	2.74	2.63	2.54	2.48	2.42	2.38	2.34	2.31
	.01	8.18	5.93	5.01	4.50	4.17	3.94	3.77	3.63	3.52	3.43	3.36	3.30
	.25	1.40	1.49	1.48	1.46	1.45	1.44	1.43	1.42	1.41	1.40	1.39	1.39
20	.10	2.97	2.59	2.38	2.25	2.16	2.09	2.04	2.00	1.96	1.94	1.92	1.89
	.05	4.35	3.49	3.10	2.87	2.71	2.60	2.51	2.45	2.39	2.35	2.31	2.28
	.01	8.10	5.85	4.94	4.43	4.10	3.87	3.70	3.56	3.46	3.37	3.29	3.23

df for Numerator

15	20	24	30	40	50	60	100	120	200	500	∞
1.53	1.52	1.52	1.51	1.51	1.50	1.50	1.49	1.49	1.49	1.48	1.48
2.24	2.20	2.18	2.16	2.13	2.12	2.11	2.09	2.08	2.07	2.06	2.06
2.85	2.77	2.74	2.70	2.66	2.64	2.62	2.59	2.58	2.56	2.55	2.54
4.56	4.41	4.33	4.25	4.17	4.12	4.08	4.01	4.00	3.96	3.93	3.91
1.50	1.49	1.49	1.48	1.47	1.47	1.47	1.46	1.46	1.46	1.45	1.45
2.17	2.12	2.10	2.08	2.05	2.04	2.03	2.00	2.00	1.99	1.98	1.97
2.72	2.65	2.61	2.57	2.53	2.51	2.49	2.46	2.45	2.43	2.42	2.40
4.25	4.10	4.02	3.94	3.86	3.81	3.78	3.71	3.69	3.66	3.62	3.60
1.48	1.47	1.46	1.45	1.45	1.44	1.44	1.43	1.43	1.43	1.42	1.42
2.10	2.06	2.04	2.01	1.99	1.97	1.96	1.94	1.93	1.92	1.91	1.90
2.62	2.54	2.51	2.47	2.43	2.40	2.38	2.35	2.34	2.32	2.31	2.30
4.01	3.86	3.78	3.70	3.62	3.57	3.54	3.47	3.45	3.41	3.38	3.36
1.46	1.45	1.44	1.43	1.42	1.42	1.42	1.41	1.41	1.40	1.40	1.40
2.05	2.01	1.98	1.96	1.93	1.92	1.90	1.88	1.88	1.86	1.85	1.85
2.53	2.46	2.42	2.38	2.34	2.31	2.30	2.26	2.25	2.23	2.22	2.21
3.82	3.66	3.59	3.51	3.43	3.38	3.34	3.27	3.25	3.22	3.19	3.17
1.44	1.43	1.42	1.41	1.41	1.40	1.40	1.39	1.39	1.39	1.38	1.38
2.01	1.96	1.94	1.91	1.89	1.87	1.86	1.83	1.83	1.82	1.80	1.80
2.46	2.39	2.35	2.31	2.27	2.24	2.22	2.19	2.18	2.16	2.14	2.13
3.66	3.51	3.43	3.35	3.27	3.22	3.18	3.11	3.09	3.06	3.03	3.00
1.43	1.41	1.41	1.40	1.39	1.39	1.38	1.38	1.37	1.37	1.36	1.36
1.97	1.92	1.90	1.87	1.85	1.83	1.82	1.79	1.79	1.77	1.76	1.76
2.40	2.33	2.29	2.25	2.20	2.18	2.16	2.12	2.11	2.10	2.08	2.07
3.52	3.37	3.29	3.21	3.13	3.08	3.05	2.98	2.96	2.92	2.89	2.87
1.41	1.40	1.39	1.38	1.37	1.37	1.36	1.36	1.35	1.35	1.34	1.34
1.94	1.89	1.87	1.84	1.81	1.79	1.78	1.76	1.75	1.74	1.73	1.72
2.35	2.28	2.24	2.19	2.15	2.12	2.11	2.07	2.06	2.04	2.02	2.01
3.41	3.26	3.18	3.10	3.02	2.97	2.93	2.86	2.84	2.81	2.78	2.75
1.40	1.39	1.38	1.37	1.36	1.35	1.35	1.34	1.34	1.34	1.33	1.33
1.91	1.86	1.84	1.81	1.78	1.76	1.75	1.73	1.72	1.71	1.69	1.69
2.31	2.23	2.19	2.15	2.10	2.08	2.06	2.02	2.01	1.99	1.97	1.96
3.31	3.16	3.08	3.00	2.92	2.87	2.83	2.76	2.75	2.71	2.68	2.65
1.39	1.38	1.37	1.36	1.35	1.34	1.34	1.33	1.33	1.32	1.32	1.32
1.89	1.84	1.81	1.78	1.75	1.74	1.72	1.70	1.69	1.68	1.67	1.66
2.27	2.19	2.15	2.11	2.06	2.04	2.02	1.98	1.97	1.95	1.93	1.92
3.23	3.08	3.00	2.92	2.84	2.78	2.75	2.68	2.66	2.62	2.59	2.57
1.38	1.37	1.36	1.35	1.34	1.33	1.33	1.32	1.32	1.31	1.31	1.30
1.86	1.81	1.79	1.76	1.73	1.71	1.70	1.67	1.67	1.65	1.64	1.63
2.23	2.16	2.11	2.07	2.03	2.00	1.98	1.94	1.93	1.91	1.89	1.88
3.15	3.00	2.92	2.84	2.76	2.71	2.67	2.60	2.58	2.55	2.51	2.49
1.37	1.36	1.35	1.34	1.33	1.33	1.32	1.31	1.31	1.30	1.30	1.29
1.84	1.79	1.77	1.74	1.71	1.69	1.68	1.65	1.64	1.63	1.62	1.61
2.20	2.12	2.08	2.04	1.99	1.97	1.95	1.91	1.90	1.88	1.86	1.84
3.09	2.94	2.86	2.78	2.69	2.64	2.61	2.54	2.52	2.48	2.44	2.42

Table 5 (continued)

df for de-nomi-nator	α	df for Numerator											
		1	2	3	4	5	6	7	8	9	10	11	12
	.25	1.40	1.48	1.47	1.45	1.44	1.42	1.41	1.40	1.39	1.39	1.38	1.37
22	.10	2.95	2.56	2.35	2.22	2.13	2.06	2.01	1.97	1.93	1.90	1.88	1.86
	.05	4.30	3.44	3.05	2.82	2.66	2.55	2.46	2.40	2.34	2.30	2.26	2.23
	.01	7.95	5.72	4.82	4.31	3.99	3.76	3.59	3.45	3.35	3.26	3.18	3.12
	.25	1.39	1.47	1.46	1.44	1.43	1.41	1.40	1.39	1.39	1.38	1.37	1.36
24	.10	2.93	2.54	2.33	2.19	2.10	2.04	1.98	1.94	1.91	1.88	1.85	1.83
	.05	4.26	3.40	3.01	2.78	2.62	2.51	2.42	2.36	2.30	2.25	2.21	2.18
	.01	7.82	5.61	4.72	4.22	3.90	3.67	3.50	3.36	3.26	3.17	3.09	3.03
	.25	1.38	1.46	1.45	1.44	1.42	1.41	1.39	1.38	1.37	1.37	1.36	1.35
26	.10	2.91	2.52	2.31	2.17	2.08	2.01	1.96	1.92	1.88	1.86	1.84	1.81
	.05	4.23	3.37	2.98	2.74	2.59	2.47	2.39	2.32	2.27	2.22	2.18	2.15
	.01	7.72	5.53	4.64	4.14	3.82	3.59	3.42	3.29	3.18	3.09	3.02	2.96
	.25	1.38	1.46	1.45	1.43	1.41	1.40	1.39	1.38	1.37	1.36	1.35	1.34
28	.10	2.89	2.50	2.29	2.16	2.06	2.00	1.94	1.90	1.87	1.84	1.81	1.79
	.05	4.20	3.34	2.95	2.71	2.56	2.45	2.36	2.29	2.24	2.19	2.15	2.12
	.01	7.64	5.45	4.57	4.07	3.75	3.53	3.36	3.23	3.12	3.03	2.96	2.90
	.25	1.38	1.45	1.44	1.42	1.41	1.39	1.38	1.37	1.36	1.35	1.35	1.34
30	.10	2.88	2.49	2.28	2.14	2.05	1.98	1.93	1.88	1.85	1.82	1.79	1.77
	.05	4.17	3.32	2.92	2.69	2.53	2.42	2.33	2.27	2.21	2.16	2.13	2.09
	.01	7.56	5.39	4.51	4.02	3.70	3.47	3.30	3.17	3.07	2.98	2.91	2.84
	.25	1.36	1.44	1.42	1.40	1.39	1.37	1.36	1.35	1.34	1.33	1.32	1.31
40	.10	2.84	2.44	2.23	2.09	2.00	1.93	1.87	1.83	1.79	1.76	1.73	1.71
	.05	4.08	3.23	2.84	2.61	2.45	2.34	2.25	2.18	2.12	2.08	2.04	2.00
	.01	7.31	5.18	4.31	3.83	3.51	3.29	3.12	2.99	2.89	2.80	2.73	2.66
	.25	1.35	1.42	1.41	1.38	1.37	1.35	1.33	1.32	1.31	1.30	1.29	1.29
60	.10	2.79	2.39	2.18	2.04	1.95	1.87	1.82	1.77	1.74	1.71	1.68	1.66
	.05	4.00	3.15	2.76	2.53	2.37	2.25	2.17	2.10	2.04	1.99	1.95	1.92
	.01	7.08	4.98	4.13	3.65	3.34	3.12	2.95	2.82	2.72	2.63	2.56	2.50
	.25	1.34	1.40	1.39	1.37	1.35	1.33	1.31	1.30	1.29	1.28	1.27	1.26
120	.10	2.75	2.35	2.13	1.99	1.90	1.82	1.77	1.72	1.68	1.65	1.62	1.60
	.05	3.92	3.07	2.68	2.45	2.29	2.17	2.09	2.02	1.96	1.91	1.87	1.83
	.01	6.85	4.79	3.95	3.48	3.17	2.96	2.79	2.66	2.56	2.47	2.40	2.34
	.25	1.33	1.39	1.38	1.36	1.34	1.32	1.31	1.29	1.28	1.27	1.26	1.25
200	.10	2.73	2.33	2.11	1.97	1.88	1.80	1.75	1.70	1.66	1.63	1.60	1.57
	.05	3.89	3.04	2.65	2.42	2.26	2.14	2.06	1.98	1.93	1.88	1.84	1.80
	.01	6.76	4.71	3.88	3.41	3.11	2.89	2.73	2.60	2.50	2.41	2.34	2.27
	.25	1.32	1.39	1.37	1.35	1.33	1.31	1.29	1.28	1.27	1.25	1.24	1.24
∞	.10	2.71	2.30	2.08	1.94	1.85	1.77	1.72	1.67	1.63	1.60	1.57	1.55
	.05	3.84	3.00	2.60	2.37	2.21	2.10	2.01	1.94	1.88	1.83	1.79	1.75
	.01	6.63	4.61	3.78	3.32	3.02	2.80	2.64	2.51	2.41	2.32	2.25	2.18

df for Numerator

15	20	24	30	40	50	60	100	120	200	500	∞
1.36	1.34	1.33	1.32	1.31	1.31	1.30	1.30	1.30	1.29	1.29	1.28
1.81	1.76	1.73	1.70	1.67	1.65	1.64	1.61	1.60	1.59	1.58	1.57
2.15	2.07	2.03	1.98	1.94	1.91	1.89	1.85	1.84	1.82	1.80	1.78
2.98	2.83	2.75	2.67	2.58	2.53	2.50	2.42	2.40	2.36	2.33	2.31
1.35	1.33	1.32	1.31	1.30	1.29	1.29	1.28	1.28	1.27	1.27	1.26
1.78	1.73	1.70	1.67	1.64	1.62	1.61	1.58	1.57	1.56	1.54	1.53
2.11	2.03	1.98	1.94	1.89	1.86	1.84	1.80	1.79	1.77	1.75	1.73
2.89	2.74	2.66	2.58	2.49	2.44	2.40	2.33	2.31	2.27	2.24	2.21
1.34	1.32	1.31	1.30	1.29	1.28	1.28	1.26	1.26	1.26	1.25	1.25
1.76	1.71	1.68	1.65	1.61	1.59	1.58	1.55	1.54	1.53	1.51	1.50
2.07	1.99	1.95	1.90	1.85	1.82	1.80	1.76	1.75	1.73	1.71	1.69
2.81	2.66	2.58	2.50	2.42	2.36	2.33	2.25	2.23	2.19	2.16	2.13
1.33	1.31	1.30	1.29	1.28	1.27	1.27	1.26	1.25	1.25	1.24	1.24
1.74	1.69	1.66	1.63	1.59	1.57	1.56	1.53	1.52	1.50	1.49	1.48
2.04	1.96	1.91	1.87	1.82	1.79	1.77	1.73	1.71	1.69	1.67	1.65
2.75	2.60	2.52	2.44	2.35	2.30	2.26	2.19	2.17	2.13	2.09	2.06
1.32	1.30	1.29	1.28	1.27	1.26	1.26	1.25	1.24	1.24	1.23	1.23
1.72	1.67	1.64	1.61	1.57	1.55	1.54	1.51	1.50	1.48	1.47	1.46
2.01	1.93	1.89	1.84	1.79	1.76	1.74	1.70	1.68	1.66	1.64	1.62
2.70	2.55	2.47	2.39	2.30	2.25	2.21	2.13	2.11	2.07	2.03	2.01
1.30	1.28	1.26	1.25	1.24	1.23	1.22	1.21	1.21	1.20	1.19	1.19
1.66	1.61	1.57	1.54	1.51	1.48	1.47	1.43	1.42	1.41	1.39	1.38
1.92	1.84	1.79	1.74	1.69	1.66	1.64	1.59	1.58	1.55	1.53	1.51
2.52	2.37	2.29	2.20	2.11	2.06	2.02	1.94	1.92	1.87	1.83	1.80
1.27	1.25	1.24	1.22	1.21	1.20	1.19	1.17	1.17	1.16	1.15	1.15
1.60	1.54	1.51	1.48	1.44	1.41	1.40	1.36	1.35	1.33	1.31	1.29
1.84	1.75	1.70	1.65	1.59	1.56	1.53	1.48	1.47	1.44	1.41	1.39
2.35	2.20	2.12	2.03	1.94	1.88	1.84	1.75	1.73	1.68	1.63	1.60
1.24	1.22	1.21	1.19	1.18	1.17	1.16	1.14	1.13	1.12	1.11	1.10
1.55	1.48	1.45	1.41	1.37	1.34	1.32	1.27	1.26	1.24	1.21	1.19
1.75	1.66	1.61	1.55	1.50	1.46	1.43	1.37	1.35	1.32	1.28	1.25
2.19	2.03	1.95	1.86	1.76	1.70	1.66	1.56	1.53	1.48	1.42	1.38
1.23	1.21	1.20	1.18	1.16	1.14	1.12	1.11	1.10	1.09	1.08	1.06
1.52	1.46	1.42	1.38	1.34	1.31	1.28	1.24	1.22	1.20	1.17	1.14
1.72	1.62	1.57	1.52	1.46	1.41	1.39	1.32	1.29	1.26	1.22	1.19
2.13	1.97	1.89	1.79	1.69	1.63	1.58	1.48	1.44	1.39	1.33	1.28
1.22	1.19	1.18	1.16	1.14	1.13	1.12	1.09	1.08	1.07	1.04	1.00
1.49	1.42	1.38	1.34	1.30	1.26	1.24	1.18	1.17	1.13	1.08	1.00
1.67	1.57	1.52	1.46	1.39	1.35	1.32	1.24	1.22	1.17	1.11	1.00
2.04	1.88	1.79	1.70	1.59	1.52	1.47	1.36	1.32	1.25	1.15	1.00